AF871316

Design, Preparation and Properties of High Performance Steels

Design, Preparation and Properties of High Performance Steels

Guest Editors

Xiucheng Li
Shilong Liu

MDPI

Basel • Beijing • Wuhan • Barcelona • Belgrade • Novi Sad • Cluj • Manchester

Guest Editors
Xiucheng Li
Collaborative Innovation
Center of Steel Technology
University of Science and
Technology Beijing
Beijing
China

Shilong Liu
School of Materials Science
and Engineering
Shanghai Jiao Tong
University
Shanghai
China

Editorial Office
MDPI AG
Grosspeteranlage 5
4052 Basel, Switzerland

This is a reprint of the Special Issue, published open access by the journal *Metals* (ISSN 2075-4701), freely accessible at: https://www.mdpi.com/journal/metals/special_issues/2OU48M269C.

For citation purposes, cite each article independently as indicated on the article page online and as indicated below:

Lastname, A.A.; Lastname, B.B. Article Title. *Journal Name* **Year**, *Volume Number*, Page Range.

ISBN 978-3-7258-8237-3 (Hbk)
ISBN 978-3-7258-8238-0 (PDF)
https://doi.org/10.3390/books978-3-7258-8238-0

© 2026 by the authors. Articles in this reprint are Open Access and distributed under the Creative Commons Attribution (CC BY) license. The reprint as a whole is distributed by MDPI under the terms and conditions of the Creative Commons Attribution-NonCommercial-NoDerivs (CC BY-NC-ND) license (https://creativecommons.org/licenses/by-nc-nd/4.0/).

Contents

About the Editors . . . . **vii**

Xiucheng Li and Shilong Liu
Design, Preparation and Properties of High-Performance Steels
Reprinted from: *Metals* **2026**, *16*, 339, https://doi.org/10.3390/met16030339 . . . . **1**

Shengfu Yuan, Ruizhi Wang, Xuelin Wang, Fajian Jiang, Chengjia Shang and Xinghua Wu
Evolution of Microstructure and the Influence of Carbides on Hardness Properties in Martensitic Stainless Steel 90Cr18MoV During Heat Treatment
Reprinted from: *Metals* **2025**, *15*, 999, https://doi.org/10.3390/met15090999 . . . . **5**

Jinghua Cong, Yongzhe Yang, Haibin Zhu, Xueliang Shang, Hongyu Wu, Zhendong Song, et al.
Study on the Coarsening Behavior of Interphase Precipitates and Random Precipitates in Steel Under the High-Temperature Environment of Fire
Reprinted from: *Metals* **2025**, *15*, 73, https://doi.org/10.3390/met15010073 . . . . **22**

Giuliano Angella, Lorenzo Ghidini and Michela Faccoli
An Alternative Method of Investigating the Thermal Stability of Shoe-Braked Railway Wheel Steels Based on Strain Hardening Analysis
Reprinted from: *Metals* **2024**, *14*, 814, https://doi.org/10.3390/met14070814 . . . . **36**

Zihan Hu and Hanguang Fu
Effect of Si Content on Microstructure and Properties of Low-Carbon Medium-Manganese Steel after Intercritical Heat Treatment
Reprinted from: *Metals* **2024**, *14*, 675, https://doi.org/10.3390/met14060675 . . . . **54**

Ba Li, Xiaoshun Zhou, Shujun Jia, Xiaoping Chen, Song Fu, Dongliang Zhao, et al.
Study on Fracture Behavior and Toughening Mechanisms of Ultra-High-Strength Pipeline Steel
Reprinted from: *Metals* **2024**, *14*, 666, https://doi.org/10.3390/met14060666 . . . . **65**

Zhan Gao, Xiaoming Dong, Jieru Yu, Junjie Tang, Yishuang Yu, Liangliang Wei, et al.
Unraveling the Mechanism of Toughness Fluctuation in Ultra-High-Strength Casing from the Perspective of Crystallography
Reprinted from: *Metals* **2024**, *14*, 208, https://doi.org/10.3390/met14020208 . . . . **86**

Jinghua Cong, Jingxiao Zhao, Xuemin Wang and Zhongwen Wu
Effect of a Gradient Temperature Rolling Process on the Microstructure and Mechanical Properties of the Center of Ultra-Heavy Plates
Reprinted from: *Metals* **2024**, *14*, 199, https://doi.org/10.3390/met14020199 . . . . **95**

En Zhu, Dianxiu Xia, Peidun Chen, Qing Han, Xuelin Wang, Zhiheng Liu and Kun Jiang
Study on the Effect of Niobium on the High Temperature Oxidation Resistance of Ferritic Stainless Steel
Reprinted from: *Metals* **2024**, *14*, 25, https://doi.org/10.3390/met14010025 . . . . **107**

SeKwon Oh, Ji-Ho Ahn, Rockhoon Jung, Hyun-Jong Kim, Younghwan Chu, Dae Hyun Choi, et al.
Characterization of Various Stainless Steels Containing Gadolinium as Thermal Neutron Absorbing and Shielding Materials
Reprinted from: *Metals* **2024**, *14*, 16, https://doi.org/10.3390/met14010016 . . . . **122**

Jingliang Wang, Rongtao Qian, Song Huang and Chengjia Shang
Effect of Double-Quenching on the Hardness and Toughness of a Wear-Resistant Steel
Reprinted from: *Metals* **2023**, *13*, 61, https://doi.org/10.3390/met13010061 . . . . **138**

About the Editors

Xiucheng Li

Xiucheng Li is an Associate Researcher at the Collaborative Innovation Center of Steel Technology, University of Science and Technology Beijing (USTB). His current research interests focus on the digital characterization of microstructures in metallic materials, the evaluation of fracture behavior and mechanical properties, as well as the application of big data and machine learning in materials science.

Shilong Liu

Shilong Liu serves as an Associate Research Fellow at the Institute of Advanced Steels and Materials, School of Materials Science and Engineering, Shanghai Jiao Tong University (SJTU). His research centers on the microstructural evolution and deformation mechanisms of advanced steels and additively manufactured metals.

metals

torial

esign, Preparation and Properties of High-Performance Steels

ucheng Li [1,*] and Shilong Liu [2,*]

1 Collaborative Innovation Center of Steel Technology, University of Science and Technology Beijing, 30 Xueyuan Road, Beijing 100083, China
2 Institute of Advanced steels and Materials, School of Materials Science and Engineering, Shanghai Jiao Tong University, 800 Dongchuan Road, Shanghai 200240, China
* Correspondence: xiuchengli@ustb.edu.cn (X.L.); sliu6@sjtu.edu.cn (S.L.)

1. Introduction

Steel remains the most widely used and consumed metal material in modern infrastructure, underpinning the continuous advancement of industrial civilization. Over the past century, the evolution of physical metallurgy has significantly enhanced steel performance. The pursuit of high-performance steel involves optimizing fundamental mechanical properties such as strength and toughness, as well as its application performance, including weldability, formability, and service life. The increasing stringency of engineering demands has driven the development of advanced steel grades with tailored properties, including corrosion resistance, fracture toughness, seismic and fire resistance, and wear resistance. These advances are strongly reliant on the advancement of physical metallurgical theory. Alloy design now considers not only solid solution and precipitation strengthening but also non-equilibrium distribution and synergistic effects among multiple elements. Microstructural control has evolved from single-phase optimization to multi-phase and multi-scale regulation, while manufacturing processes have grown increasingly precise, requiring tighter processing windows. Driven by continued innovation in metallurgical theory, coupled with advances in industrial technology and equipment, steel performance will undoubtedly be enhanced further, resulting in materials that can meet the complex challenges of future engineering systems. This Special Issue provides a forum for the high-performance-steel research community, reporting new developments and innovations in this field.

2. Contributions

Jingliang Wang et al. (contribution 1) investigated the effect of a double-quenching process on a wear-resistant steel. This treatment refined the austenite grain size by 43%, leading to a simultaneous improvement in both toughness and hardness: the ductile-to-brittle transition temperature decreased from −77 °C to −90 °C, while hardness increased by 8%. The findings also suggest that the improved properties are primarily attributed to the prior austenite grain boundaries and high-angle packet boundaries, whose effectiveness is governed by the high misorientation angle between the {110} slip planes and a high slip transmission factor.

SeKwon Oh et al. (contribution 2) investigated austenitic (304, 316) and duplex (5A) stainless steels reinforced with 2 wt.% gadolinium (Gd) for thermal neutron absorption in spent nuclear fuel applications. The evenly distributed Gd intermetallics slightly increased yield and tensile strength but reduced elongation and impact toughness. Corrosion resistance also diminished. Despite these trade-offs, the Gd-alloyed steels demonstrated superior neutron absorption capacity compared to commercial-grade borated stainless steel (ASTM A887). Fractography revealed the intermetallics' influence on fracture behavior. The findings suggest that Gd-reinforced stainless steels are promising candidates for spent nuclear fuel storage and transportation, offering enhanced thermal neutron shielding despite moderate mechanical property degradation.

En Zhu et al. (contribution 3) investigated the effect of Nb content on the high-temperature oxidation resistance of ferritic stainless steel for automotive exhaust applications. The study revealed that a higher Nb content promotes the precipitation of fine NbN at grain boundaries, refining the microstructure and enhancing oxidation resistance, particularly at temperatures up to 900 °C. These findings demonstrate that optimizing the Nb content is critical for extending exhaust pipe service life by improving oxide layer stability at elevated temperatures.

Jinghua Cong et al. (contribution 4) investigated the effect of gradient temperature rolling (GTR) on the center properties of ultra-heavy plates. Compared to traditional uniform temperature rolling, GTR significantly increased central deformation, refining prior austenite grains and increasing grain boundary densities (prior austenite grain boundaries by ~140%, high-angle packet boundaries by ~39%, and low-angle packet boundaries by ~49%). This microstructural refinement enhanced strength, plasticity, and low-temperature toughness, and the impact fracture mode transitioned from semi-brittle to ductile. These findings demonstrate that GTR effectively mitigates center property degradation in ultra-heavy plates by promoting grain refinement and increasing boundary density, which are the key factors governing mechanical performance.

Zhan Gao et al. (contribution 5) investigated the mechanism of toughness fluctuation in ultra-high-strength casing from a crystallographic perspective. The study revealed that uneven cooling along the casing leads to microstructural variations: the head, which has a higher cooling rate, consists of lath bainite with a higher density of high-angle block boundaries, while the end, which has a lower cooling rate, contains mixed granular and lath bainite with fewer block boundaries. Consequently, the head exhibits 41% higher impact toughness than the end. These findings demonstrate that cooling rate-induced variant selection governs block boundary density, which is the key crystallographic factor determining toughness uniformity in ultra-high-strength casing.

Ba Li et al. (contribution 6) investigated the fracture behavior and toughening mechanisms of ultra-high-strength pipeline steel through low-temperature Charpy V-notch impact test (CVN) and drop-weight tear test (DWTT) experiments. The results showed that DWTT fractures exhibited more complex morphological features than CVN fractures, better reflecting the actual fracturing process. Microstructural analysis revealed that inclusion of fine-grained granular bainite enhanced toughness by inducing crack deflections and frequent small transitions, which consumed additional energy. In contrast, large-grained granular bainite with low-angle grain boundaries failed to effectively hinder crack propagation, and the aggregation of M/A constituents promoted microcrack formation, impairing toughness.

Zihan Hu et al. (contribution 7) investigated the influence of Si content on the microstructure and mechanical properties of a low-carbon medium-manganese steel subjected to intercritical heat treatment. The results show that with increasing Si content, the fraction of ferrite decreases, while both the ferrite and martensite constituents become finer and more uniformly distributed. This refinement and homogenization of the microstructure enhances mechanical performance. The optimal comprehensive properties were achieved at a Si content of 0.685 wt.%, yielding a tensile strength of 726 MPa, an elongation of 30.5%, and a strength–ductility product of 22,143 MPa·%.

Giuliano Angella et al. (contribution 8) investigated the tensile work hardening behavior of five railway wheel steels by interpolating flow curves using the Voce constitutive

https://doi.org/10.3390/met160303

equation, which correlates its parameters with microstructural features based on dislocation density evolution. Their analysis confirmed a strong correlation between Voce parameters, microstructural characteristics, and apparent fracture toughness. Furthermore, by evaluating performance changes and microstructural evolution after heat treatments at various temperatures, this work demonstrates the effectiveness of work hardening analysis in assessing microstructural stability and changes in the mechanical properties of railway wheel steels subjected to thermal cycles during service.

Jinghua Cong et al. (contribution 9) investigated the coarsening of interphase and random precipitates in fire-resistant steel through a novel heat treatment protocol designed to achieve both precipitation types. The results show that the growth and coarsening rates of interphase precipitates are substantially lower those of than random precipitates. The superior thermal stability of interphase precipitates is attributed to their excellent size uniformity, which inhibits concentration gradient formation. After prolonged tempering, interphase precipitates maintained higher strengthening contributions. These findings provide novel insights and theoretical foundations for the design and development of fire-resistant steels.

Shengfu Yuan et al. (contribution 10) investigated the effect of quenching and tempering parameters on 90Cr18MoV martensitic stainless steel. Using a Thermo-Calc optimized quenching and tempering heat treatment, a maximum microhardness and nanohardness of 57 HRC and 7.2 GPa were achieved due to secondary hardening. Microstructural analysis revealed that secondary hardening arose from the co-precipitation of $M_{23}C_6$ and MC carbides within the matrix. Carbides $\leq$20 nm enhance matrix hardness, while those exceeding 30 nm lead to a reduction in hardness after tempering. The findings provide novel insights into the effects of precipitate size on mechanical performance.

3. Conclusions and Outlook

The contributions in this Special Issue highlight the critical role of physical metallurgy principles in tailoring the mechanical properties and service reliability of steel materials. Through controlled thermomechanical processing, precipitation engineering, and alloy design, the ten studies demonstrate how microstructural features—grain boundaries, precipitates, and phases—can be tailored to achieve application-specific performance. Spanning railway wheel steels, neutron-absorbing stainless steels, ultra-high-strength pipeline steels, and fire-resistant steels, this collection offers reference cases for heat treatment dynamics, alloying element effects, precipitate behavior, and the fundamental process–structure–property relationships that ultimately govern mechanical behavior and service reliability.

Building on these insights, future research on high-performance steels will emphasize application-specific performance metrics beyond basic strength and toughness, establishing quantitative microstructure–property relationships tailored to service conditions. The integration of materials informatics and artificial intelligence will accelerate alloy design and process optimization, enabling predictive modeling of complex material behaviors. These approaches will facilitate the development of next-generation steels with optimized performance for targeted applications.

Acknowledgments: As Guest Editors, we are very grateful to all the reviewers and authors who contributed to this Special Issue, as well as to the Editorial staff for their dedicated support throughout the entire publication process.

Conflicts of Interest: The author declares no conflicts of interest.

https://doi.org/10.3390/met16030339

List of Contributions:

1. Wang, J.; Qian, R.; Huang, S.; Shang, C. Effect of Double-Quenching on the Hardness and Toughness of a Wear-Resistant Steel. *Metals* **2023**, *13*, 61. https://doi.org/10.3390/met13010061.
2. Oh, S.; Ahn, J.-H.; Jung, R.; Kim, H.-J.; Chu, Y.; Choi, D.H.; Lee, H.; Jung, H.-D. Characterization of Various Stainless Steels Containing Gadolinium as Thermal Neutron Absorbing and Shielding Materials. *Metals* **2024**, *14*, 16. https://doi.org/10.3390/met14010016.
3. Zhu, E.; Xia, D.; Chen, P.; Han, Q.; Wang, X.; Liu, Z.; Jiang, K. Study on the Effect of Niobium on the High Temperature Oxidation Resistance of Ferritic Stainless Steel. *Metals* **2024**, *14*, 25. https://doi.org/10.3390/met14010025.
4. Cong, J.; Zhao, J.; Wang, X.; Wu, Z. Effect of a Gradient Temperature Rolling Process on the Microstructure and Mechanical Properties of the Center of Ultra-Heavy Plates. *Metals* **2024**, *14*, 199. https://doi.org/10.3390/met14020199.
5. Gao, Z.; Dong, X.; Yu, J.; Tang, J.; Yu, Y.; Wei, L.; Liu, S.; Jin, X. Unraveling the Mechanism of Toughness Fluctuation in Ultra-High-Strength Casing from the Perspective of Crystallography. *Metals* **2024**, *14*, 208. https://doi.org/10.3390/met14020208.
6. Li, B.; Zhou, X.; Jia, S.; Chen, X.; Fu, S.; Zhao, D.; Zhang, H.; Guo, J. Study on Fracture Behavior and Toughening Mechanisms of Ultra-High-Strength Pipeline Steel. *Metals* **2024**, *14*, 666. https://doi.org/10.3390/met14060666.
7. Hu, Z.; Fu, H. Effect of Si Content on Microstructure and Properties of Low-Carbon Medium-Manganese Steel after Intercritical Heat Treatment. *Metals* **2024**, *14*, 675. https://doi.org/10.3390/met14060675.
8. Angella, G.; Ghidini, L.; Faccoli, M. An Alternative Method of Investigating the Thermal Stability of Shoe-Braked Railway Wheel Steels Based on Strain Hardening Analysis. *Metals* **2024**, *14*, 814. https://doi.org/10.3390/met14070814.
9. Cong, J.; Yang, Y.; Zhu, H.; Shang, X.; Wu, H.; Song, Z.; Wang, X.; Xu, X. Study on the Coarsening Behavior of Interphase Precipitates and Random Precipitates in Steel Under the High-Temperature Environment of Fire. Metals **2025**, *15*, 73. https://doi.org/10.3390/met15010073.
10. Yuan, S.; Wang, R.; Wang, X.; Jiang, F.; Shang, C.; Wu, X. Evolution of Microstructure and the Influence of Carbides on Hardness Properties in Martensitic Stainless Steel 90Cr18MoV During Heat Treatment. *Metals* **2025**, *15*, 999. https://doi.org/10.3390/met15090999.

Disclaimer/Publisher's Note: The statements, opinions and data contained in all publications are solely those of the individu author(s) and contributor(s) and not of MDPI and/or the editor(s). MDPI and/or the editor(s) disclaim responsibility for any injury people or property resulting from any ideas, methods, instructions or products referred to in the content.

https://doi.org/10.3390/met160303

metals

MDPI

icle

volution of Microstructure and the Influence of Carbides on ardness Properties in Martensitic Stainless Steel 90Cr18MoV uring Heat Treatment

engfu Yuan [1,2], Ruizhi Wang [1,2], Xuelin Wang [2,3], Fajian Jiang [2], Chengjia Shang [2,4] and Xinghua Wu [1,2,*]

[1] School of Materials and Energy, Guangdong University of Technology, Guangzhou 510006, China; yuansf_77@163.com (S.Y.); wrzyjhj@163.com (R.W.)
[2] Yangjiang Branch, Guangdong Laboratory for Materials Science and Technology (Yangjiang Advanced Alloys Laboratory), Yangjiang 529500, China; xuelin2076@ustb.edu.cn (X.W.); 13602235170@163.com (F.J.); cjshang@ustb.edu.cn (C.S.)
[3] Collaborative Innovation Center of Steel Technology, University of Science and Technology Beijing, Beijing 100083, China
[4] State Key Laboratory for Advanced Metals and Materials, University of Science and Technology Beijing, Beijing 100083, China
* Correspondence: wuxi0012@e.ntu.edu.sg

Abstract: In this study, we utilized Thermo-Calc software (2023a) to optimize the heat treatment process of martensitic stainless steel 90Cr18MoV through phase diagram calculations. The microhardness of 90Cr18MoV was characterized using a nanoindentation instrument. The microstructural morphology of the samples was analyzed using scanning electron microscopy (SEM). The composition of the samples was characterized through scanning electron backscatter diffraction (EBSD) and X-ray diffraction (XRD). Additionally, laser confocal microscopy (FIB) and transmission electron microscopy (TEM) were employed to characterize the precipitate phase composition and size before and after heat treatment, while also observing the dislocation structure within the samples. The relationship between the quenching temperature and the percentage of residual austenite content in the material was established. The influence of the dislocation structure and precipitate size on the hardness of the samples was investigated. The research findings confirm that the observed secondary hardening phenomenon in tempered samples is attributed to the co-precipitation of two types of carbides, $M_{23}C_6$ and MC, within the matrix. The study investigated the effects of the tempering temperature and duration on the size of secondary precipitates, indicating that $M_{23}C_6$ and MC particles with sizes less than or equal to 20 nm contribute to enhancing the matrix, while particles larger than 30 nm lead to a reduction in hardness after tempering. Notably, during the tempering process, $M_{23}C_6$ precipitated from the matrix nucleates on MC.

Keywords: heat treatment; hardness; secondary precipitates; $M_{23}C_6$; MC

1. Introduction

The high-carbon, high-chromium martensitic stainless steel 90Cr18 is widely used in the production of high-end bearings, surgical instruments, and valves due to its exceptional hardness and wear resistance [1]. The addition of alloying elements in high-carbon martensitic stainless steels can obstruct slip planes, thereby enhancing their mechanical properties [2]. Furthermore, an increase in the chromium content present significantly improves the material's corrosion resistance [3,4]. Due to the strong reducibility of chromium

(Cr), it is widely used in stainless steel formulations [5]. Similarly, the addition of a small proportion of molybdenum (Mo) has been shown to improve the resistance of stainless steel against pitting and crevice corrosion in chloride environments [6].

The alloy 90Cr18MoV is classified as a high-carbon, high-chromium martensitic stainless steel. Martensite forms as a supersaturated solid solution of carbon (C) within alpha iron (α-Fe), making it easier for high-carbon steels to develop a robust martensitic structure [7]. During the melting and solidification stages of 90Cr18 martensitic stainless steel, there is a tendency for the formation of difficult-to-remove network or plate-like carbides such as M_7C_3 and $M_{23}C_6$; "M" mainly refers to Cr [8]. The formation of large-sized carbides like Cr_7C_3 and $Cr_{23}C_6$ inevitably leads to regions deficient in chromium surrounding these carbides, which significantly diminishes the localized corrosion resistance of 90Cr18 martensitic stainless steel. Moreover, atoms such as manganese (M) or vanadium (V) can substitute chromium atoms at positions within $Cr_{23}C_6$. This substitution releases Cr from the carbides into the matrix material, which greatly enhances localized corrosion resistance while also facilitating dislocation accumulation that contributes positively to increasing the material strength [9].

Early studies have indicated that the size and composition of M_xC_y carbides are related to the cooling rate, and an excessively large carbide size significantly affects material properties [10,11]. It is well known that heat treatment methods can facilitate the phase transformation from M_7C_3 to $M_{23}C_6$ [10]. The phase transformation may initiate either from the center of primary carbide M_7C_3 [12] or from the peripheral regions of M_7C_3 [10,13,14]. Wang et al. [15] discovered that in high-performance microalloyed (HP-MA) steels, the transition from M_7C_3 to $M_{23}C_6$ at a heat treatment temperature of 1000 °C (Quenching) is controlled by the diffusion of carbon from the shell of $M_{23}C_6$ into the matrix. Yu et al. [16] investigated carbide evolution in high-carbon 8Cr13MoV stainless steel and found it challenging to eliminate the formed network carbides during solidification under conventional heat treatment and thermal deformation techniques. However, thermal deformation proves to be effective in fragmenting network carbides to smaller sizes by controlling deformation temperatures [16]. The thermal deformation at lower temperatures than Ac1 yields better fragmentation results [17]. Nevertheless, the higher concentrations of C and Cr in 90Cr18 make it more feasible for the formation of coarse network carbides, leading to an increased hardness that could not to be worked cold. Consequently, the current processing of 90Cr18MoV martensitic stainless steel predominantly employs hot rolling techniques. Currently, the temperature of hot rolling processes is lower than 800 °C for energy saving, which could not completely dissolve and fragment these network carbides. Cracking inevitably occurs, requiring further toughening treatments. To investigate suitable heat treatment processes, Qian et al. [18] carried out spheroidizing annealing and quenching-tempering processes on 60Cr16MoMA martensitic stainless steels and found that the coarse, fragmented M_7C_3 carbides transformed into $M_{23}C_6$ carbides.

However, it was found that post-heat-treated 60Cr16MoMA did not eliminate the detrimental effects of coarse $M_{23}C_6$ on the material; instead, an increase in the size of M23C6 was observed after tempering, indicating a lack of precise control over carbide dimensions. The carbide is inevitably formed in high-carbon martensitic stainless steels. Large-sized carbides lead to decreased corrosion resistance and increased brittleness in materials. In comparison, fine carbides tend to enhance material properties such as the presence of nanoscale chromium-rich $M_{23}C_6$ carbides that enhance the hardness of the microstructure [19].

In recent years, there has been growing interest among researchers regarding the size and type of secondary precipitated carbides due to their beneficial role in strengthening the matrix [20]. Li et al. [21] indicated that large-sized $M_{23}C_6$ carbides with sizes ranging from 0.3 μm to 1.5 μm significantly improve low-temperature toughness in martensitic heat-resistant steels. Bonagani et al. [22] conducted a tempering process on the steels containing 13%wt. Cr at temperatures above 550 °C and found that the Cr-depleted zones with the sizes of approximately 7–9 nm appeared around the secondary precipitated carbides, resulting in a decline in the pitting corrosion resistance of the materials.

Currently, research on high-carbon martensitic stainless-steel carbides primarily focuses on the types of primary carbides formed in the martensitic matrix after heat treatment. However, there remains a lack of comprehensive studies regarding the types of secondary carbides and the size variations of these carbides at different temperatures. In this study, the influence of different quenching temperatures on the phase transformation of 90Cr18MoV was investigated, as well as the effect of the residual austenite content resulting from quenching on the hardness of the matrix and the whole materials. After quenching and tempering treatment, the mechanism of secondary hardening in 90Cr18MoV during the tempering process was investigated. By analyzing the size of secondary phase precipitates generated during tempering, we studied the influence of carbide dimensions on the hardness of the 90Cr18MoV microstructure. Additionally, we examined the relationship between the size of secondary precipitates and the tempering temperature. Furthermore, an analysis was conducted on the precipitation and formation mechanisms of $M_{23}C_6$.

2. Materials and Methods

2.1. Materials

The material utilized in this study is martensitic stainless-steel plate 90Cr18MoV, which has undergone hot rolling and annealing processes(This material has been independently smelted and rolled by our research institution.). The chemical composition of the steels is shown in Table 1. The sample size is 10 mm × 8 mm × 4 mm (length × width × height). The heat treatment process for 90Cr18MoV stainless steel is illustrated in Figure 1. As shown in Figure 1, this experiment investigates the quenching treatment of 90Cr18MoV within a temperature range of 1020 °C to 1100 °C, with increments of 20 °C as individual research subjects. The quenching duration is set at 20 min, and oil is utilized for cooling the samples. To explore the effects of different tempering temperatures on the internal structure and properties of the material, we examine a temperature range from 150 °C to 700 °C, stepping in increments of 50 °C as separate research subjects. The tempering time is established at 180 min. We designate the material after quenching as Q-W, where Q represents the quenching process and W denotes the quenching temperature. We designated the samples that underwent tempering as QT-X/Y, where QT denotes quenching followed by tempering, X represents the tempering temperature, and Y indicates the duration of tempering.

Table 1. Chemical composition of the 90Cr18MoV steel (=wt. %).

C	Si	Mn	P	S	Cr	Ni	Mo	V
0.903	0.512	0.420	0.026	0.004	17.19	0.329	1.069	0.161

To investigate the microstructure of the obtained samples, the samples after heat treatment were polished and then etched with a mixture of 90 mL of 4% ferric chloride solution and 10 mL of hydrochloric acid for 5 s. After that, an electrolyte polishing process was performed on the samples at a voltage of 30 V and a current of 1 mA for electron backscatter diffraction analysis. The electrolyte solution contained 10% perchloric acid and 90% alcohol to obtain specimens suitable for EBSD analysis.

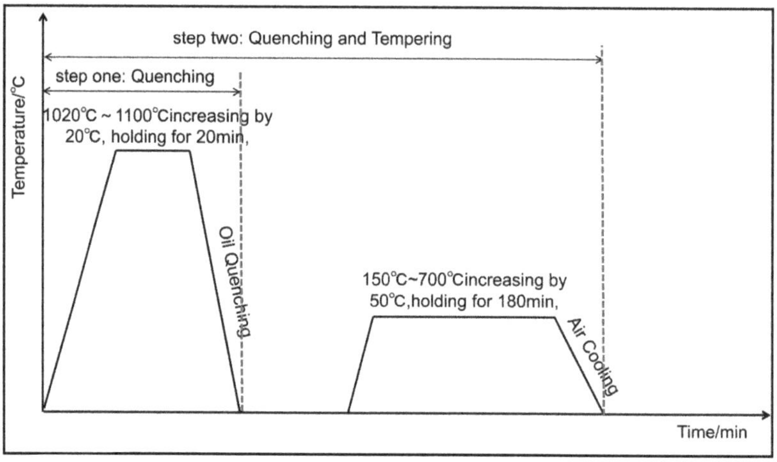

Figure 1. The optimized heat treatment process of 90Cr18MoV.

2.2. Characterization

The microstructure of 90Cr18MoV was observed using Leica-DMI8C optical microscopy (OM) (Wetzlar, Germany) and a TESCAN/CLARA GMH (Brno, Czech Republic) field emission scanning electron microscope (FESEM). An Oxford Symmetry S2 electron backscatter diffraction system (EBSD) and Oxford UltimMax65x energy-dispersive X-ray spectroscopy (EDS) (Abingdon, UK), in conjunction with a Thermo Scientific Talos F200X S/TEM (Waltham, MA, USA) transmission electron microscope (STEM), were used for electron microscopy observation. Transmission specimens were prepared utilizing a dual-beam thinning instrument and Helios 5UX focused ion beam (FIB) (Waltham, MA, USA) technology.

Nano-hardness and elastic modulus characterization of the samples were performed using a Bruker Nano Inc. (Billerica, MA, USA). TI Premier nanoindentation tester, applying a load of 10 mN.

3. Results

3.1. Quenching Treatment

Figure 2a shows the morphology of the microstructure of the 90Cr18MoV samples as received. The carbides (orange marker) occupied a large portion of the 90Cr18MoV matrix. The corresponding TEM images of carbides are shown in Figure 2b–d. The diffraction patterns indicate that these large carbides are composed mainly of M_7C_3 inside grains (marked in yellow) while the small carbides are composed mainly of $M_{23}C_6$ (marked in blue).

Martensite formed due to the post-quenching processes, as a supersaturated solid solution of carbon, has been believed to present high hardness [23]. To obtain high hardness, the 90Cr18MoV were quenched at different temperatures of 1020 °C, 1040 °C, 1060 °C, 1080 °C, and 1100 °C. We designated these samples as Q-1020, Q-1040, Q-1060, Q-1080, and Q-1100, respectively. The microstructure of 90Cr18MoV before and after different quenching temperatures are shown in Figure 3. The presence of a large area of precipitated carbides in the original sample, as shown in Figure 3a, is attributed to the annealing process that facilitates the precipitation within the carbide matrix, thereby softening the material and making it easier to process.

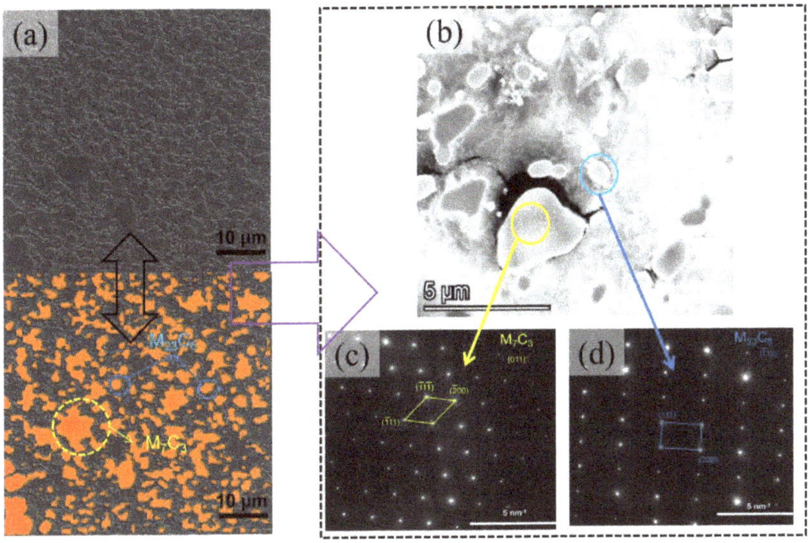

Figure 2. (**a**) The morphology of the microstructure the 90Cr18MoV samples as received.; (**b–d**) The corresponding TEM image and diffraction patterns of carbides.

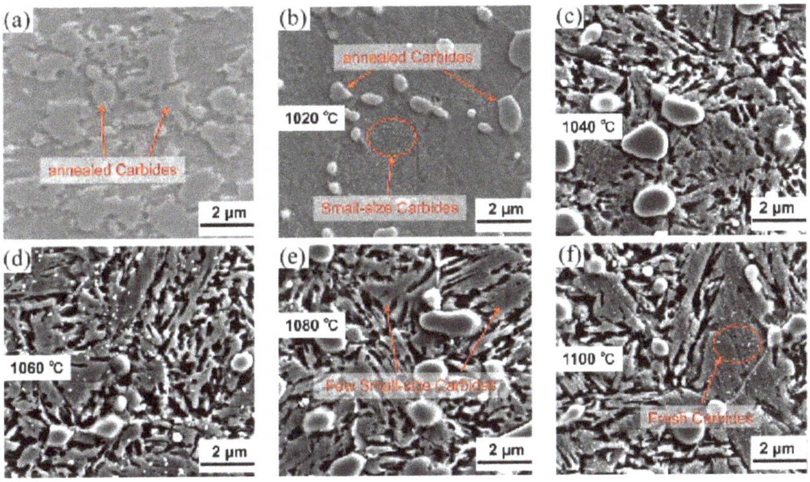

Figure 3. SEM images of 90Cr18MoV after quenching. Panels (**a–f**) represent the microstructural morphology at quenching temperatures of Untreated, Q-1020, Q-1040, Q-1060, Q-1080, and Q-1100, respectively.

As the temperature increases, the carbides in the matrix gradually diminish. Larger carbides decrease in size, while smaller carbides completely dissolve back into the matrix. In Figure 3e, it can be observed that no minute carbides are present in the matrix of the Q-1080 sample. However, as shown in Figure 3f, with a further elevation of the temperature, carbides begin to reappear in the matrix of the Q-1100 sample. At the same time, an increase in carbon concentration of the matrix resulted in a decrease in the martensitic transformation temperature. This phenomenon leads to an incomplete austenite to martensite transformation under different quenching conditions. The phase structure analysis was conducted on the samples subjected to quenching using EBSD, and the results are presented in Figure 4. At a quenching temperature of 1040 °C, the retained austenite having an fcc structure is labeled in red and its area fraction is 0.5% as shown in Figure 4a–d. The fraction of the fcc structure increased to 1.15% after quenching at a temperature of 1060 °C. The fraction of the fcc structure increased to 6.42% after quenching at a temperature of 1080 °C. After quenching at 1100 °C, the fraction of the fcc structure reached up to 35.8%. Therefore, as the quenching temperature increases, the content of austenite in the matrix also rises. Concurrently, fine carbides gradually dissolve back into the matrix with increasing quenching temperatures, leading to an increase in carbon content within the matrix. Since carbon (C) is a stabilizing element for austenite [24], a higher carbon content enhances the stability of austenite during heat treatment. Consequently, residual austenite becomes more prevalent with rising quenching temperatures, which aligns with findings from other studies [25,26]. In this research, we determined that the critical quenching temperature for carbon dissolution is 1060 °C.

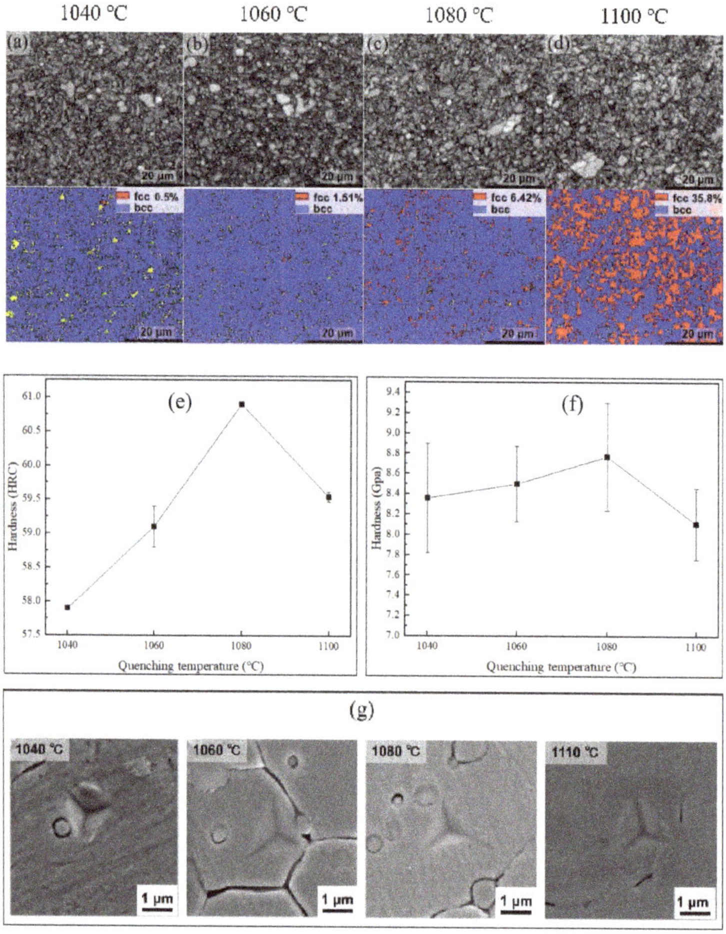

Figure 4. EBSD analysis of 90Cr18MoV samples quenched at different temperatures and their corresponding hardness: (**a**) EBSD image of the Q-1060; (**b**) EBSD image of the Q-1080; (**c**) EBSD image of the Q-1100; (**d**) Macrohardness of samples quenched at various temperatures; (**e**) Nanoindentation hardness of samples subjected to quenching at different temperatures; (**f**) SEM scanning images showing hardness points from nanoindentation on the matrix for samples quenched at varying temperatures; (**g**) Nano Hardness Position Scanning Image.

To investigate the effect of an increasing quenching temperature on the microstructure and hardness of 90Cr18MoV, we conducted hardness characterization on samples quenched at different temperatures, as illustrated in Figure 4d,e. Under identical quenching times, the macroscopic hardness of 90Cr18MoV initially increases with a rising quenching temperature before subsequently decreasing. A similar trend is observed in the microscopic organizational hardness, which also shows an initial increase followed by a decline. Notably, samples quenched at 1080 °C exhibit maximum hardness values of 61 HRC and 8.8 GPa. The hardness of 440C martensitic stainless steel obtained through deep cryogenic treatment, as reported by A. Idayan et al., is approximately 60 HRC [27]. To validate that the measured microscopic hardness reflects the material's structural integrity, we performed SEM scanning analysis on samples post-nanoindentation testing, as shown in Figure 4f.

3.2. *Tempering Treatment*

To eliminate the internal structural defects of samples after rolling and quenching, it is necessary to perform tempering treatment on the quenched samples. As shown in Figure 4, the hardness of samples quenched at a temperature of 1060 °C is the highest. Additionally, from the EBSD results of the quenched samples, it can be observed that there is minimal residual austenite present when quenched at this temperature. Therefore, we will perform wire cutting on the Q-1060 samples to divide them into multiple smaller specimens for subsequent tempering treatment.

As illustrated in Figure 5, SEM scanning images of tempered samples at different temperatures reveal that the microstructure primarily consists of martensite and some dispersed carbides. Furthermore, with an increase in the tempering temperature, a significant number of precipitates appears within the matrix structure. Specifically, as depicted in Figure 5a, numerous precipitates are observed in the matrices of samples tempered at temperatures of 550 °C and 600 °C, respectively. Zhenfeng Xu et al. [28] conducted a study on the precipitation behavior of $M_{23}C_6$ under various boundary conditions, confirming that $M_{23}C_6$ precipitates at a temperature of 600 °C.

To investigate the relationship between the hardness of tempered samples and tempering temperature, we conducted a macro-hardness analysis on samples tempered at different temperatures, as illustrated in Figure 5b. The hardness of the samples exhibited an initial increase followed by a decrease with rising tempering temperatures. Additionally, to examine the secondary hardening characteristics of the matrix in tempered samples, we performed nanoindentation hardness analysis on specimens tempered at 460 °C, 480 °C, 500 °C, and 520 °C. By combining nanoindentation techniques with scanning electron microscopy (SEM), we assessed the hardness of the matrix for these specific tempering temperatures. As shown in Figure 5c,d the maximum nanoindentation hardness achieved during secondary hardening of the tempered samples was found to be 7.2 GPa. The maximum macrohardness is 57 HRC.

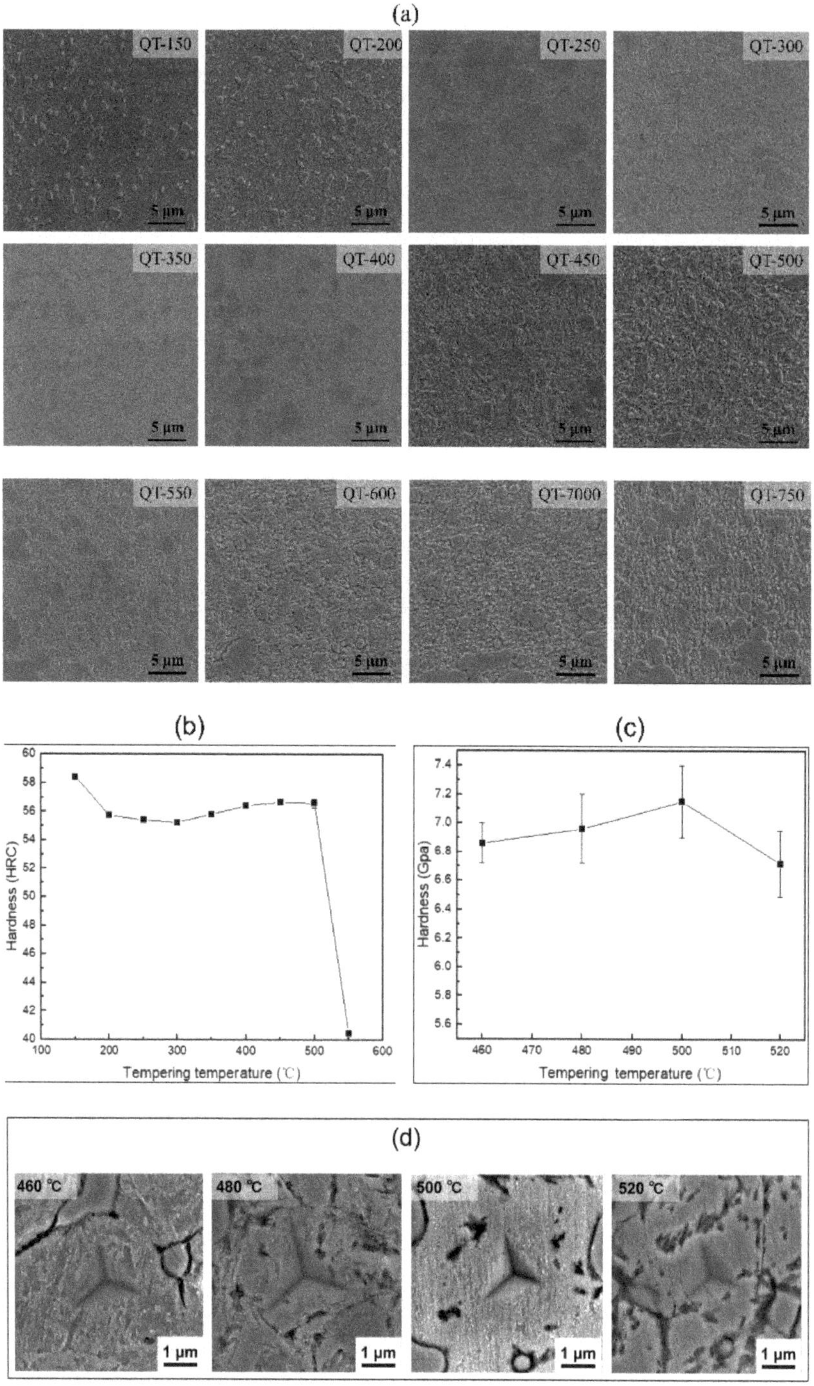

Figure 5. Analysis of the microstructure and hardness characterization of experimental steel; (**a**) Scanning images of samples at different tempering temperatures, (**b**,**c**) Trends in hardness variation of samples at different tempering temperatures, (**d**) Scanning images of selected regions for nanoindentation.

4. Discussion

4.1. Phase Diagram Calculation

The phase diagram was calculated using Thermo-Calc software, as illustrated in Figure 6a. The high-temperature liquid phase exists above 1250 °C, while the austenite phase is present within the temperature range of 800 °C to 1380 °C. In this study, the quenching temperatures employed are 1020 °C, 1040 °C, 1060 °C, 1080 °C, and 1100 °C; Thus, these quenching temperatures fall within the austenite phase region. Furthermore, with an increase in the quenching temperature, the relative content of the austenite phase also gradually increases, which is consistent with the EBSD results presented in Figure 4. The bcc structure appears between temperatures of 1350 °C and 1420 °C, as well as below temperatures of 830 °C. The two-phase region comprising fcc and bcc structures exists within the temperature range of 800 °C to 850 °C. Therefore, all tempering processes conducted in this study occur outside this two-phase region. $M_{23}C_6$ can be found at temperatures below 1180 °C while M_7C_3 may appear within a temperature range from 1150 °C to 1270 °C.

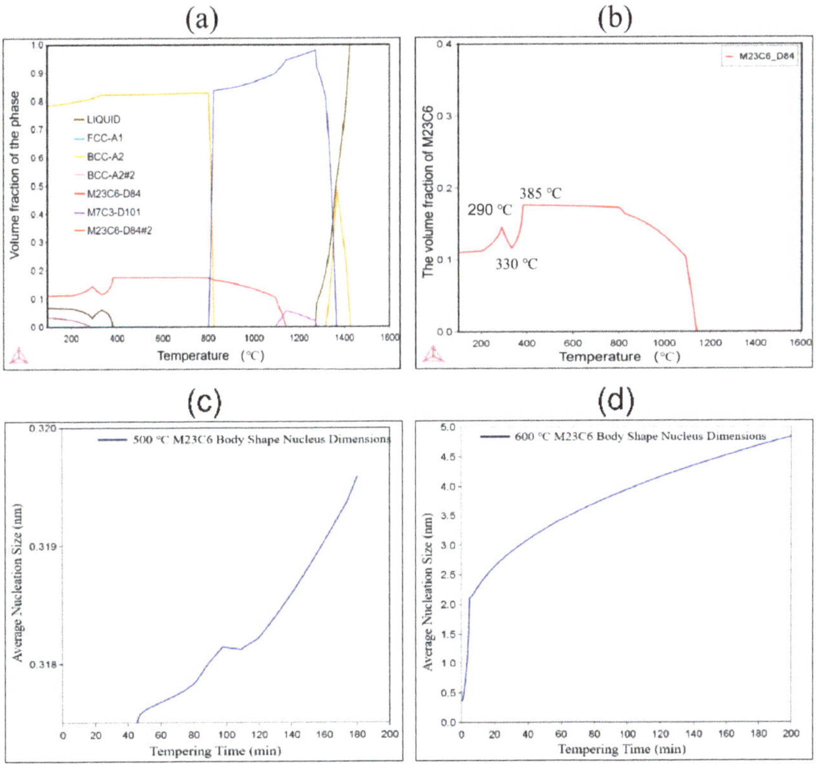

Figure 6. (**a**) The volume fractions of various phases in 90Cr18MoV steel; (**b**) Simulation calculation of $M_{23}C_6$ phase content; (**c**) Precipitation model of $M_{23}C_6$ in 90Cr18MoV at complete equilibrium state of 500 °C; (**d**) Precipitation model of $M_{23}C_6$ in 90Cr18MoV at complete equilibrium state of 600 °C.

Figure 6b presents the phase composition diagram of $M_{23}C_6$ at equilibrium, simulated using Thermo-Calc. It is evident that there is a significant increase in the presence of $M_{23}C_6$ within the temperature range of 330 °C to 385 °C, after which it begins to re-dissolve into the matrix at temperatures exceeding 800 °C. Although the processing methods employed in this study do not allow for the samples to achieve complete equilibrium at identical temperatures, the overall trend remains consistent. Consequently, during the tempering process, precipitation of $M_{23}C_6$ will occur. In order to investigate the precipitation conditions of the $M_{23}C_6$ phase that are compatible with industrial production, we will employ transmission electron microscopy for a more detailed analysis of the samples.

To investigate the precipitation models of $M_{23}C_6$ and MC at different temperatures, this study employs Thermo-Calc to analyze and simulate the nucleation and growth of $M_{23}C_6$ and MC in samples using both bulk nucleation and grain boundary nucleation models. As shown in Figure 6c, under isothermal conditions at 500 °C, simulations were conducted to assess the nucleation and growth of $M_{23}C_6$ over varying holding times in a fully equilibrated state. Similarly, Figure 6d presents simulations for the same parameters at 600 °C.

It can be observed that at a temperature of 500 °C, as the holding time increases, both $M_{23}C_6$ exhibit continuous growth in their nucleation sizes. Likewise, at a temperature of 600 °C, an increase in the holding time also leads to an enhancement in the size of $M_{23}C_6$ nuclei. However, it is noteworthy that the sizes of precipitated carbides ($M_{23}C_6$) formed at 600 °C are significantly greater than those formed at 500 °C.

4.2. Microstructural Characterization

In order to investigate the reasons for the changes in hardness of samples under different tempering conditions, we conducted STEM analysis on samples tempered at various temperatures, as illustrated in Figure 7. Figure 7a–c represent the tempered samples at temperatures of 150 °C, 250 °C, and 350 °C, respectively. It can be seen that the number of dislocations formed after quenching significantly decreased with an increase in the tempering temperature. Zhang et al. [29] investigated the toughening mechanisms during the tempering process of high chromium steel, revealing that as the tempering temperature increases, the dislocation structure within the experimental steel significantly diminishes. Combining this observation with the relationship curve between the sample hardness and tempering temperature shown in Figure 5, we conclude that higher tempering temperatures effectively reduce structural defects; consequently, as these organizational defects diminish, the material hardness also tends to decrease. It is evident that as the tempering temperature increases, the number of residual structural defects (dislocations) remaining after quenching significantly decreases. This microstructural morphology corroborates the relationship curve between sample hardness and tempering temperature presented in Figure 5.

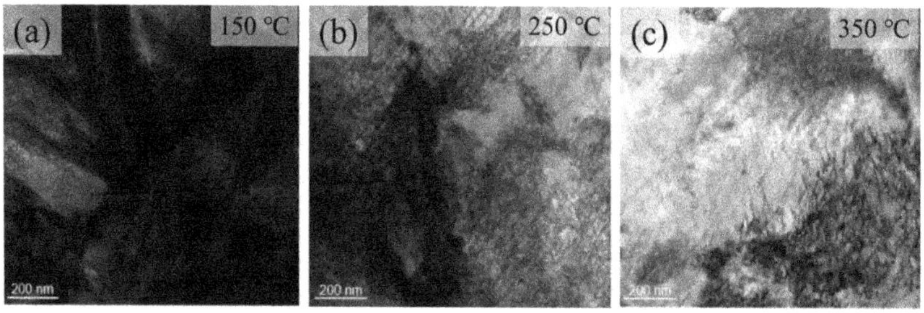

Figure 7. Transmission electron microscopy analysis of samples QT-150, QT-250, and QT-350: (**a**) Dark field image of the QT-150 sample; (**b**) Dark field image of the QT-250 sample; (**c**) Dark field image of the QT-350 sample.

However, a continuous increase in the tempering temperature resulted in the formation of nanoscale precipitates within the matrix of the 90Cr18MoV alloy as shown in Figure 8a–c. Wang et al. [30] conducted a study on the aging strengthening of high-carbon high-Cr steel and found that the carbides precipitated during the tempering process have a certain reinforcing effect on the matrix. Furthermore, accumulation of dislocations around these fine carbides was observed, leading to a secondary hardening effect observed in the tempered samples. To gain a deeper understanding of the specific characteristics of these precipitates, further detailed analyses are required.

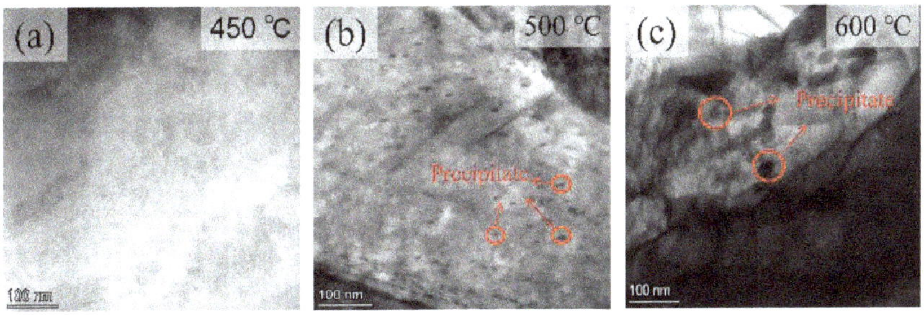

Figure 8. (**a**) Transmission image of the QT-450/180 sample; (**b**) Transmission image of the QT-500/180 sample; (**c**) Transmission image of the QT-600/180 sample.

To further investigate the lattice structures of the above samples, we initially employed Electron Backscatter Diffraction (EBSD) to study the lattice structure of the samples post-tempering; the results are shown in Figure 9. Figure 9a presents the QT-500/180. The results indicate the presence of smaller carbides surrounding larger carbides. These are small carbides, specifically MoC and VC; some researchers have experimentally verified that carbides such as VC precipitate during the tempering process [31,32]. They are observed to precipitate exclusively around the particles of the larger carbides. The tempered samples were scanned by XRD and are shown in Figure 9c. The results again verified the existence of MoC and VC in the matrix after quenching. Furthermore, it was noted that the content of $M_{23}C_6$ and MC increased continuously with rising tempering temperatures.

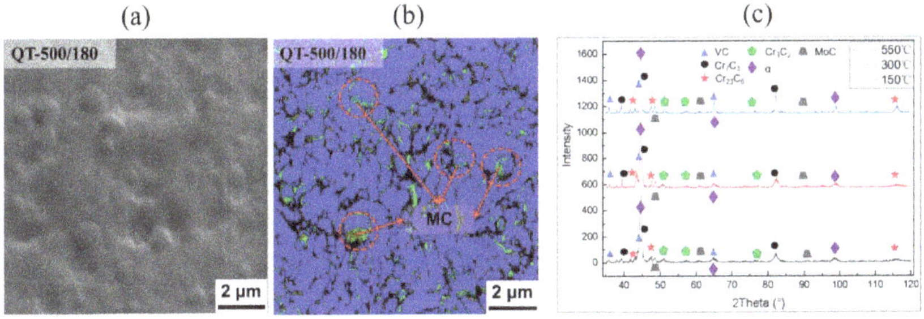

Figure 9. EBSD and XRD analyses of the tempered samples: (**a**,**b**) present the EBSD images of the QT-500/180 sample; (**c**) illustrates the XRD analysis for QT-150/180, QT-300/180, and QT-550/180 samples.

To verify the specific structure and composition of carbides precipitated during high-temperature tempering, TEM diffraction analysis was conducted on the carbide precipitation regions. FIB was employed to cut and thin the tempered samples as shown in Figure 10a,b. Figure 10c presents the QT-450/180 in which no fine carbides were observed within the matrix. In contrast, Figure 10d illustrates QT-500/180 where the average size of the precipitated carbides is approximately 20 nm. The corresponding diffraction pattern is shown in Figure 10e. It indicates that the carbides are VC and $M_{23}C_6$. The positional correlation of the diffraction spots indicates that $M_{23}C_6$ is nucleated on MC. Figure 10f illustrates the transmission dark field image of the precipitates in the QT-600/180 sample. The corresponding diffraction pattern is shown in Figure 10g, and the carbides were verified to be $M_{23}C_6$ and the average size of the precipitated carbides is 50 nm as shown in Figure 10f. Figure 10j shows QT-700/180 with the transmission dark field image. The corresponding diffraction pattern is shown in Figure 10k, and the carbides were verified to be $M_{23}C_6$ while the average size larger than 100 nm is shown in Figure 10j. These small $M_{23}C_6$ directly precipitate from the matrix during post-tempering and grow gradually with the increasing tempering temperature.

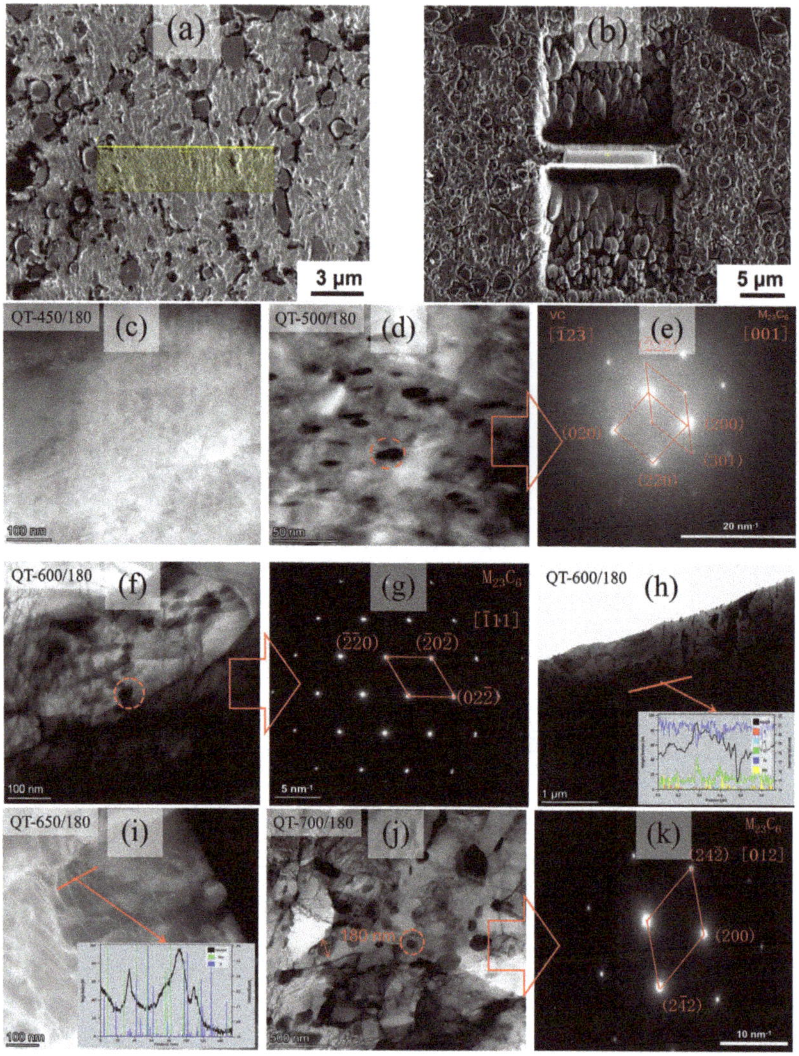

Figure 10. (**a**,**b**) illustrate transmission samples prepared using a focused ion beam (FIB); (**c**) presents the transmission image of the QT-450/180 sample; (**d**) shows the transmission image of the QT-500/180 sample; (**e**) displays the diffraction pattern corresponding to the QT-500/180 sample at a specific location; (**f**) depicts the transmission image of the QT-600/180 sample; (**g**) provides the diffraction pattern for a specific location in the QT-600/180 sample; (**h**) offers an energy-dispersive spectroscopy (EDS) line scan for the QT-600/180 sample; (**i**) presents an EDS line scan for the QT-650/180 sample; (**j**) illustrates the transmission image of the QT-700/180 sample; and finally, (**k**) shows the diffraction pattern corresponding to a specific location in the QT-700/180 sample.

4.3. Refinement of Precipitate Size Statistics

The carbides precipitated during tempering are randomly dispersed within the matrix, contributing to dispersion strengthening. From Figure 10, it can be observed that the precipitated carbides exhibit an fcc structure corresponding to M23C6. To quantitatively investigate the effect of carbide size on the hardness performance of 90Cr18MoV due to tempering, we employed transmission electron microscopy (TEM) to statistically analyze the sizes of precipitates in samples tempered at different temperatures, as illustrated in Figures 10j, 11b and 12d,f. As the tempering temperature increases, the average size of the carbides gradually rises. Specifically, after tempering for 180 min at a temperature of 500 °C, the average size of precipitates is measured at 19.9 nm. At a temperature of 600 °C, the average size of carbides in tempered samples reaches 54.7 nm; while at a temperature of 700 °C, this value further increases to an average size of 152.86 nm. The overall trend is consistent with the simulation calculation illustrated in Figure 6.

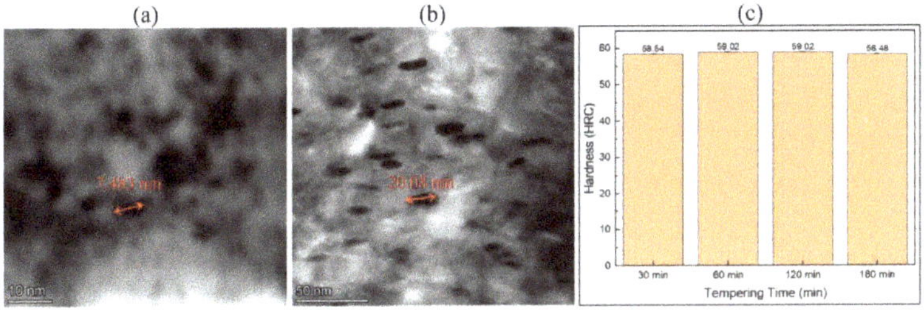

Figure 11. (**a**) Statistical analysis of precipitate sizes for the QT-500/60 sample; (**b**) Statistical analysis of precipitate sizes for the QT-500/180 sample; (**c**) Correlation between tempering hardness and tempering time for the Q-1080 sample at a temperature of 500 °C.

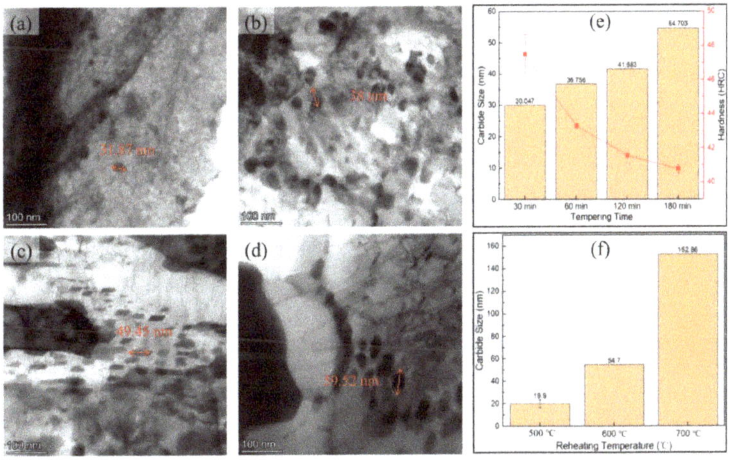

Figure 12. (**a**) Size distribution statistics of precipitates in the QT-600/30 sample; (**b**) Size distribution statistics of precipitates in the QT-600/60 sample; (**c**) Size distribution statistics of precipitates in the QT-600/120 sample; (**d**) Size distribution statistics of precipitates in the QT-600/180 sample; (**e**) Statistical analysis of the average size of precipitates and its relationship with tempering time, as well as the correlation between sample hardness and tempering time for the QT-600 samples; (**f**) Size distribution statistics of precipitates for samples QT-500/180, QT-600/180, and QT-700/180.

At the same time, to investigate the effect of different tempering durations on the size of secondary precipitated carbides at a constant temperature, we conducted tempering experiments on samples that were quenched at 1080 °C for various durations (30 min,

60 min, 120 min, and 180 min) at a temperature of 600 °C. The results are illustrated in Figure 12a–e. The average size of the precipitates after tempering for 30 min at 600 °C was approximately 30 nm. After tempering for 60 min, the average size increased to about 38 nm; following a duration of 120 min, it reached around 41 nm; and after tempering for an extended period of 180 min, the average size further increased to approximately 54 nm. As shown in Figure 12e it is evident that with an increasing tempering time, there is a corresponding decrease in the sample hardness.

In the process of studying the effect of different tempering times at the same temperature on the size of precipitated carbides in samples, we observed that at a tempering temperature of 500 °C, the sizes of precipitates varied with different tempering durations, as illustrated in Figure 11a,b. The size of precipitates after 60 min of tempering was approximately 10 nm, while after 180 min it increased to around 20 nm. However, despite the increase in tempering time, there was no significant change in sample hardness within the margin of error; as shown in Figure 11c, all samples maintained a hardness level around 59 HRC. By correlating Figures 10j, 11b and 12d, we conclude that when the tempering temperature exceeds 500 °C, an increase in this temperature leads to continuous growth of secondary precipitate $M_{23}C_6$ within the sample matrix. When $M_{23}C_6$ is sized at or below approximately 20 nm, it contributes positively to enhancing the sample hardness. Conversely, once $M_{23}C_6$ exceeds a size greater than 30 nm, its further growth results in a declining trend in the sample hardness.

4.4. The Formation Mechanism of $M_{23}C_6$

The presence of $M_{23}C_6$ in the tempered samples leads us to conclude that, in the samples tempered at 500 °C, we observed the formation of carbides with sizes of 20 nm and below. The TEM images of the QT-450/180 samples revealed that such fine carbides were not present in the matrix (see Figure 10d), Therefore, it is certain that the fine carbides precipitated at 500 °C are formed because of the diffusion of C and Cr elements within the matrix. Additionally, the presence of VC was identified within these small carbides as shown in Figure 9. The XRD and EBSD analyses conducted on different tempering samples indicate that MC is present in the tempered samples, with its quantity increasing alongside the rise in the tempering temperature. Furthermore, the analysis of diffraction spots presented in Figure 10e demonstrates that the secondary precipitation of $M_{23}C_6$ during tempering occurs at MC as a nucleation point. The schematic diagram illustrating the nucleation of $M_{23}C_6$ on MC is shown in Figure 13. Furthermore, this precipitate phase is beneficial for facilitating the nucleation of the fine carbides within the matrix [33]. The combination of V with C occupies the sites that were originally held by Cr and C pairs. The released Cr elements deceased the amount of precipitated chromium-containing carbides and affected the corrosion resistance of samples. Moreover, the growth behavior of these carbide precipitates is influenced by elemental diffusion.

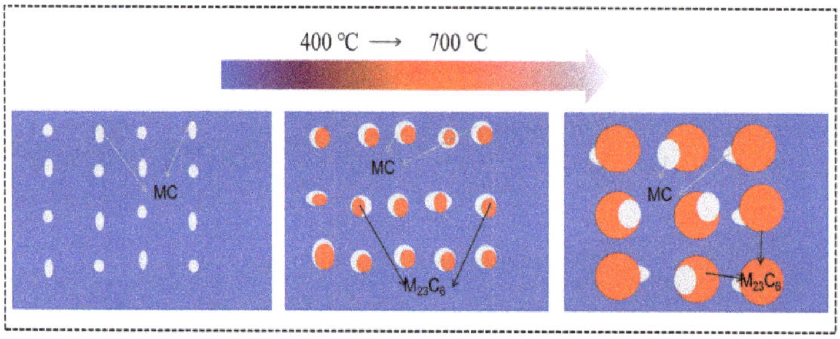

Figure 13. The precipitation process of $M_{23}C_6$ in the matrix during tempering.

5. Conclusions

This study investigated the effects of quenching and tempering temperatures and durations on the microstructure and mechanical properties of 90Cr18MoV, leading to the following conclusions.

(1) The 90Cr18MoV alloy, quenched at a temperature of 1080 °C, presented a small amount of retained austenites. The increase in the carbon content within the matrix contributes to the increase in hardness. The measured microhardness and nano-hardness were 61 HRC and 8.8 GPa, respectively.

(2) The effect of the tempering temperature on the quenched 90Cr18MoV samples at a temperature of 1080 °C was investigated. A maximum microhardness and nanohardness of 57 HRC and 7.2 GPa were achieved at the tempering temperature of 500 °C due to the secondary hardening.

(3) The phenomenon of secondary hardening during tempering is attributed to the combined effects of the precipitation of MC and $M_{23}C_6$ carbides, which are smaller than 20 nm in size, within the matrix. The $M_{23}C_6$ that precipitates during tempering serves as a nucleation site for further nucleation processes occurring in the matrix, originating from MC.

(4) At a tempering temperature of 500 °C, maintaining the process for 180 min can result in the formation of secondary precipitates approximately 20 nm in size, which enhances the hardness of the microstructure. In contrast, secondary precipitates larger than 30 nm tend to reduce the hardness of the microstructural features.

Author Contributions: Conceptualization, S.Y. and X.W. (Xinghua Wu); methodology, S.Y. and X.W. (Xuelin Wang); software, F.J.; validation, R.W., F.J. and C.S.; formal analysis, S.Y., R.W. and X.W. (Xinghua Wu); investigation, R.W., X.W. (Xuelin Wang) and X.W. (Xinghua Wu); resources, C.S.; data curation, S.Y., R.W., F.J. and X.W. (Xinghua Wu); writing—original draft preparation, R.W.; writing—review and editing, S.Y. and X.W. (Xinghua Wu); visualization, F.J. and C.S.; supervision, X.W. (Xuelin Wang) and C.S.; funding acquisition, X.W. (Xuelin Wang) and C.S. All authors have read and agreed to the published version of the manuscript.

Funding: This research was funded by the program "Research of Warm Rolling Process for High Carbon Martensitic Stainless Steel and Powder Steel Used in High end Knife and Scissors, and Promotion and Application of New Materials" (SDZX202204), and this work was supported by the Yangjiang Advanced Alloys Laboratory through its Independent Research Project (Grant No. YJAAKY2025-001).

Data Availability Statement: The original contributions presented in this study are included in the article. Further inquiries can be directed to the corresponding authors.

Conflicts of Interest: The authors declare no conflicts of interest.

ferences

Wang, X.L.; Huang, A.R.; Shang, C.J.; Xie, Z.J. Characterization of the cladding layer by laser cladding of 9Cr18Mo powder on 3Cr14 martensitic stainless steel and the impact of martensite obtained through post heat treatment on hardness. *Mater. Today Commun.* **2022**, *32*, 104057. [CrossRef]

Mul, D.O.; Bushueva, E.G.; Lazurenko, D.V.; Lozhkina, E.A.; Domarov, E.V. Structure and tribological properties of "carbon steel-VC containing coating" compositions formed by non-vacuum electron-beam surfacing of vanadium-containing powder mixtures. *Surf. Coat. Technol.* **2023**, *474*, 130107. [CrossRef]

Chen, T.; Hao, L.; Liu, T.; Zhong, Y.; Wang, Z.; Liu, C.; Cheng, X.; Li, X. Insights into the role of the Cr and rare element in improving the corrosion resistance of HRB400 rebars in simulated SO_2-polluted marine environment. *J. Build. Eng.* **2024**, *97*, 110807. [CrossRef]

Liu, L.; Lu, K.; Jing, Z.; Zhang, Z.; Xue, L.; Yuan, J.; Li, X.; Zhang, B.; Liang, X.; Cheng, J. Simultaneously improving strength and corrosion resistance of CoCrNi-based multi-principal element alloys via alloying with Mo content. *J. Alloys Compd.* **2025**, *1010*, 177119. [CrossRef]

5. Hao, Y.; Xu, R.; Bi, H.; Zhang, Z.; Chen, Z.; Li, M.; Chen, B. The corrosion properties of ferritic stainless steel with varying Cr a Mo contents in the early stages of a simulated proton exchange membrane fuel cell environment investigated using experimen and joint calculation method HKD. *Corros. Sci.* **2024**, *239*, 112389. [CrossRef]
6. Liu, Z.B.; Ling, J.-X.; Yang, Z.Y. Effect and application of Mo in martensitic precipitation hardening stainless steel. *Contin. C* **2015**, *34*, 44–48.
7. Krauss, G. Martensite in steel: Strength and structure. *Mater. Sci. Eng. A* **1999**, *273–275*, 40–57.
8. Halfa, H.; Mattar, T.; Eissa, M. Precipitation behavior of modified as cast high nitrogen super hard high-speed tool steel. *Steel R* **2012**, *83*, 1071–1078. [CrossRef]
9. Yue, X.; Chen, D.; Krishnan, A.; Lazar, I.; Niu, Y.; Golias, E.; Wiemann, C.; Gloskovskii, A.; Schlueter, C.; Jeromin, A.; et Unveiling nano-scale chemical inhomogeneity in surface oxide films formed on V- and N-containing martensite stainless steel synchrotron X-ray photoelectron emission spectroscopy/microscopy and microscopic X-ray absorption spectroscopy. *J. Ma Sci. Technol.* **2025**, *205*, 191–203. [CrossRef]
10. Wieczerzak, K.; Bala, P.; Dziurka, R.; Tokarski, T.; Cios, G.; Koziel, T.; Gondek, L. The effect of temperature on the evolution eutectic carbides and $M_7C_3 \rightarrow M_{23}C_6$ carbides reaction in the rapidly solidified Fe-Cr-C alloy. *J. Alloys Compd.* **2017**, *698*, 673–6 [CrossRef]
11. Kong, F.; Zhao, W.; Li, H.; He, N. Effect of grain size and cobalt content on machining performance during milling tungst carbide with PCD tool. *Int. J. Refract. Met. Hard Mater.* **2024**, *123*, 106780. [CrossRef]
12. Kondrat'ev, S.Y.; Kraposhin, V.S.; Anastasiadi, G.P.; Talis, A.L. Experimental observation and crystallographic description M_7C_3 carbide transformation in Fe-Cr-Ni-C HP type alloy. *Acta Mater.* **2015**, *100*, 275–281. [CrossRef]
13. Wang, K.; Li, D. Formation of core (M_7C_3)-shell ($M_{23}C_6$) structured carbides in white cast irons: A thermo-kinetic analys *Comput. Mater. Sci.* **2018**, *154*, 111–121.
14. Wiengmoon, A.; Chairuangsri, T.; Brown, A.; Brydson, R.; Edmonds, D.V.; Pearce, J.T.H. Microstructural and crystallographi study of carbides in 30wt.%Cr cast irons. *Acta Mater.* **2005**, *53*, 4143–4154.
15. Wang, M.; Flahaut, D.; Zhang, Z.; Jones, I.P.; Chiu, Y.L. Primary carbide transformation in a high performance micro-alloy 1000 °C. *J. Alloys Compd.* **2019**, *781*, 751–760.
16. Yu, W.T. Study on Control Technology of Carbides in High-Carbon Martensitic Stainless Steel 8Cr13MoV Used as Knives a Shears. Ph.D. Thesis, University of Science and Technology Beijing, Beijing, China, 2017.
17. Zuo, J.Z.; He, X.; Zhao, Y.; Zhou, Y.; Chen, T.J. Influence of hot deformation process on network carbide of 100Cr6 bearing st wire. *Spec. Steel* **2022**, *43*, 60–65. [CrossRef]
18. Qian, S.; Teng, H.; Zhao, H.; Hu, P.; Man, T.; Dong, H. Microstructural evolution of secondary carbides during spheroidiz annealing and quenching and tempering in 60Cr16MoMA martensitic stainless steel. *J. Mater. Res. Technol.* **2024**, *28*, 3207–32 [CrossRef]
19. Lu, S.Y.; Yao, K.F.; Chen, Y.B.; Wang, M.H.; Ge, X.Y. Influence of heat treatment on the microstructure and corrosion resistance 13 Wt Pct Cr-type martensitic stainless steel. *Metall. Mater. Trans. A* **2015**, *46*, 6090–6102. [CrossRef]
20. Muchiri, P.W.; Mwalukuku, V.M.; Korir, K.K.; Amolo, G.O.; Makau, N.W. Hardness characterization parameters of niobiu carbide and niobium nitride: A first principles study. *Mater. Chem. Phys.* **2019**, *229*, 489–494. [CrossRef]
21. Li, J.; Zhang, C.; Jiang, B.; Zhou, L.; Liu, Y. Effect of large-size $M_{23}C_6$-type carbides on the low-temperature toughness martensitic heat-resistant steels. *J. Alloys Compd.* **2016**, *685*, 248–257. [CrossRef]
22. Bonagania, S.K.; Bathulab, V.; Kain, V. Influence of tempering treatment on microstructure and pitting corrosion of 13 wt.% martensitic stainless steel. *Corros. Sci.* **2018**, *131*, 340–354. [CrossRef]
23. Lu, Y.; Yu, H.; Sisson, R.D., Jr. The effect of carbon content on the c/a ratio of as-quenched martensite in Fe-C alloys. *Mater. S Eng. A* **2017**, *700*, 592–597. [CrossRef]
24. Zhang, J.; Xia, D.X.; Xu, R.C.; Sun, Z.H.; Wu, S.S.; Zhang, X.T.; Chen, T. The Genetic Influence of Carbon Segregation in GCr Cast Billets on Microstructure and Friction-Wear Properties. *Steel* **2025**, 1–14. [CrossRef]
25. Luo, Q.; Chen, H.; Chen, W.; Wang, C.; Xu, W.; Li, Q. Thermodynamic prediction of martensitic transformation temperature Fe-Ni-C system. *Scr. Mater.* **2020**, *187*, 413–417. [CrossRef]
26. Navarrete-Cuadrado, J.; Soria-Biurrun, T.; Lozada-Cabezas, L.; Isasti, N.; Ibarreta-López, F.; Martínez-Pampliega, R.; Sánch Moreno, J.M. Effect of carbon content and cooling rate on the microstructure and hardness of TiC-Fe-Cr-Mo cermets. *Int. J. Refra Met. Hard Mater.* **2024**, *119*, 106552. [CrossRef]
27. Idayan, A.; Gnanavelbabu, A.; Rajkumar, K. Influence of deep cryogenic treatment on the mechanical properties of AISI 44 bearing steel. *Procedia Eng.* **2014**, *97*, 1683–1691.
28. Xu, Z.; Xia, Z.; Xiong, Y.; Lu, J.; Guo, Z. Multiple morphologies and twin structure generated by a definite growth behavior grain boundary $M_{23}C_6$ in tempered Fe–15Mn–3Al-0.7C steel. *J. Mater. Res. Technol.* **2024**, *28*, 774–781.
29. Zhang, J.; Yu, L.; Liu, C.; Ding, R.; Liu, Y. Synergistic Strengthening Mechanisms and Deformation Heat Treatment Applicatio of High Chromium Martensitic Heat-Resistant Steel. *Acta Metall. Sin.* **2024**, *60*, 713–730.

Wang, X.; Xu, L.; Jiao, L.; Li, W.; Mei, J.; Zhao, Y.; Qiao, L. Inhibition of the intergranular brittleness of HR3C heat-resistant steel by strain-aging induced nano-$M_{23}C_6$ dispersion precipitation. *J. Mater. Sci. Technol.* **2025**, *213*, 288–299.
Wang, X.; Li, C.; Zhang, Z.; Zhao, X.; Tong, D.; Yan, G.; Han, L.; Gu, J. Tailoring the microstructure and mechanical properties of H13 steel by controlling the pre-precipitation of VC carbides from austenite. *J. Mater. Res. Technol.* **2024**, *33*, 9614–9621.
Wang, Q.; Kong, D.; Li, X.; Zhou, S.; Zhang, Z. Additive manufacturing Cr-Mo-Si-V steel: Systematic parameter assessments, precipitation behavior of in-situ VC-$M_{23}C_6$ and strengthening mechanisms. *Mater. Sci. Eng. A* **2025**, *919*, 147504. [CrossRef]
Jiao, Y.; Dan, W.J.; Zhang, W.G. The strain-induced martensitic phase transformation of Fe–C alloys considering C addition: A molecular dynamics study. *J. Mater. Res.* **2020**, *35*, 1803–1816. [CrossRef]

sclaimer/Publisher's Note: The statements, opinions and data contained in all publications are solely those of the individual hor(s) and contributor(s) and not of MDPI and/or the editor(s). MDPI and/or the editor(s) disclaim responsibility for any injury to ple or property resulting from any ideas, methods, instructions or products referred to in the content.

metals

Article

Study on the Coarsening Behavior of Interphase Precipitates and Random Precipitates in Steel Under the High-Temperature Environment of Fire

Jinghua Cong [1,†], Yongzhe Yang [1,†], Haibin Zhu [1], Xueliang Shang [1], Hongyu Wu [1], Zhendong Song [2], Xuemin Wang [1,*] and Xiangyu Xu [3,*]

1 Collaborative Innovation Center of Steel Technology, University of Science and Technology Beijing, Beijing 100083, China; congjinghua@126.com (J.C.); yangyongzhe0812@163.com (Y.Y.); haibinzhu123@163.com (H.Z.); shangxl1996@163.com (X.S.); whymomei@sina.cn (H.W.)
2 Technical Center of Inner Mongolia Baotou Steel Union Co., Ltd., Baotou 014010, China; szhd1982@126.com
3 Center for Advanced Solidification Technology, School of Materials Science and Engineering, Shanghai University, Shanghai 200444, China
* Correspondence: wxm@mater.ustb.edu.cn (X.W.); xuxiangyu@shu.edu.cn (X.X.)
† These authors contributed equally to this work.

Abstract: In the domain of fire-resistant steels, the characteristics of precipitates significantly influence material properties. This study developed a novel heat treatment protocol to concurrently achieve both interphase precipitation and random precipitation. Samples were subjected to isothermal treatments at various temperatures and durations, while techniques such as scanning electron microscopy (SEM) and transmission electron microscopy (TEM) were employed to thoroughly analyze the coarsening behavior of the two types of precipitate and reveal their thermal stability differences. The results show that the growth and coarsening rates of interphase precipitates are substantially lower than random precipitates. Coarsening kinetics analysis reveals that the radius of random precipitates follows a 1/3 power law with time at 600 °C and 650 °C, whereas the radius of interphase precipitates adheres to a 1/6 power law at 600 °C and a 1/5 power law at 650 °C. Furthermore, interphase precipitation demonstrates excellent size uniformity, which hinders the formation of a concentration gradient, thereby reducing the coarsening rate and enhancing thermal stability. After prolonged tempering treatment, interphase precipitation maintains a higher strengthening contribution than random precipitation. This study provides novel insights and theoretical foundations for the design and development of fire-resistant steels.

Keywords: interphase precipitation; random precipitation; thermal stability; coarsening behavior

1. Introduction

The behavior of precipitates in steels plays a crucial role in determining the mechanical properties, high-temperature performance, and long-term durability of these materials [1–3]. These steels are widely used in demanding applications such as power plants, high-temperature reactors, and turbine components, where their ability to withstand elevated temperatures without significant degradation is critical [4–6]. Studies have shown that precipitate formation and coarsening during heat treatment processes are pivotal in optimizing steel properties by influencing the microstructure and controlling mechanisms such as creep resistance, oxidation resistance, and thermal stability [7–9]. Precipitation hardening in fire-resistant steels, or other steels requiring high-temperature performance,

https://doi.org/10.3390/met150100

is typically achieved through the formation of fine, stable precipitates, such as carbides, nitrides, and intermetallic compounds. However, prolonged exposure to high temperatures leads to the coarsening of these precipitates, which can result in a reduction in the steel's strength and resistance to deformation.

The coarsening of precipitates is a thermally activated process involving atom migration within the precipitate matrix, driven by differences in chemical potential between the precipitate and the surrounding matrix [6–9]. Understanding the mechanisms behind precipitate coarsening is essential for optimizing the performance of fire-resistant steels and other steels requiring high-temperature performance. Higher temperatures generally accelerate coarsening, and this relationship has been extensively studied in fire-resistant steels, where service conditions typically involve operation at elevated temperatures for extended periods. Studies have shown that the rate of precipitate coarsening depends on several factors, including temperature, alloying elements, and the initial size distribution of precipitates [10–13]. Wang [8] and Dudko [9] find that controlled heat treatments can significantly reduce precipitate coarsening, thereby enhancing the steel's strength at elevated temperatures. Furthermore, precipitate coarsening plays a crucial role in the creep behavior of these steels. Kesternich [14] and Miyata [15] find that creep resistance is closely linked to the size and distribution of precipitates, with finer precipitates effectively pinning grain boundaries and dislocations, thereby improving resistance to deformation under high-temperature stress. Therefore, a deeper understanding of the coarsening process and its influencing factors can inform the design of steels with superior performance in high-temperature environments.

Interphase precipitation, as a special type of precipitation, has garnered particular interest from researchers [16–18]. The contribution of interphase precipitation to strength at room temperature can reach approximately ~300 MPa [19]. Additionally, Yang's research finds that interphase precipitation plays a significant role in creep strengthening [20]. While substantial progress has been made in understanding the mechanisms of interphase precipitation and its role in strengthening, the high-temperature stability of these precipitates under service conditions remains largely underexplored. The stability of interphase precipitates at elevated temperatures is critical, as their coarsening or dissolution can result in the loss of the beneficial strengthening effect, particularly during long-term high-temperature exposure. Recent studies on precipitate coarsening have primarily focused on the bulk behavior of precipitates in alloys, without distinguishing between the precipitates formed during the phase transformation process and those formed in the bulk matrix. In contrast, the high-temperature coarsening behavior of interphase precipitates remains inadequately investigated. We have recently noted that interphase precipitation can effectively enhance the fire resistance of steels, primarily through precipitation strengthening and dislocation strengthening at elevated temperatures [21]. The aforementioned studies on interphase precipitation at elevated temperatures indirectly suggest that interphase precipitation exhibits excellent thermal stability.

In the present investigation, to investigate the thermal stability of interphase precipitation, a novel heat treatment method was developed to simultaneously induce both interphase and random precipitation at varying temperatures. Following isothermal treatments at different temperatures and durations, the coarsening behaviors and size uniformity of the two types of precipitate were analyzed to investigate the differences in their thermal stability and the underlying reason. The results of this study could provide an alternative approach to the design and development of fire-resistant steels.

2. Materials and Methods

Table 1 shows the chemical composition of the experimental steel measured using Inductively Coupled Plasma (ICP) and Inert Gas Fusion (IGF). The experimental steel was cast and rolled in the form of plate with a thickness of 14 mm. The experimental steel samples were homogenized in a quartz tube at 1200 °C for 72 h. Next, they were quenched in water. Thermo-mechanical treatments were then carried out using a continuous annealing simulator (ULVAC, Osaka, Japan). These heat treatment schedules are shown schematically in Figure 1. The sample after austenitizing at 950 °C and isothermal heat treatment at 650 °C is hereafter referred to as S0. The sample held at 600 °C for 1 h is hereafter referred to as S0-600 °C-1 h. Similarly, S0-600 °C-3 h and S0-600 °C-10 h are also named. The samples subjected to isothermal holding at 650 °C for different durations were labeled as S0-650 °C-1 h, S0-650 °C-3 h and S0-650 °C-10 h, respectively. SEM (scanning electron microscope) micrographs and TEM (transmission electron microscopy) micrographs were used to observe the precipitates. Specimens for microstructural studies were conducted using 3 mm diameter thin foils and mechanically thinned to 0.05 mm and electropolished using a solution of 5% perchloric acid and 95% alcohol at −25 °C. Then, a part of them was etched with 3% nital solution and observed using a JSM-7001F field emission scanning electron microscope (JEOL Ltd., Tokyo, Japan). To avoid statistical errors caused by thickness variations, SEM micrographs were used for the statistical analysis of the average diameter and density of precipitates, with a minimum of 20 images being analyzed in each specimen. Another part was used for confirmation of precipitate types by JEM-2100 TEM (JEOL Ltd., Tokyo, Japan). MATLAB software (R2021b) was used to programmatically process the SEM images of the precipitates, making them more accessible for observation, statistical analysis, and further interpretation. Microhardness measurements of ferrite were taken randomly using a Vickers hardness tester (CAS, Shanghai, China) with a load of 0.01 kg.

Table 1. The chemical composition of the experimental steel (wt.%).

C	Mn	Si	Ti	S	P	Fe
0.07	1.30	0.18	0.10	0.009	0.012	Bal.

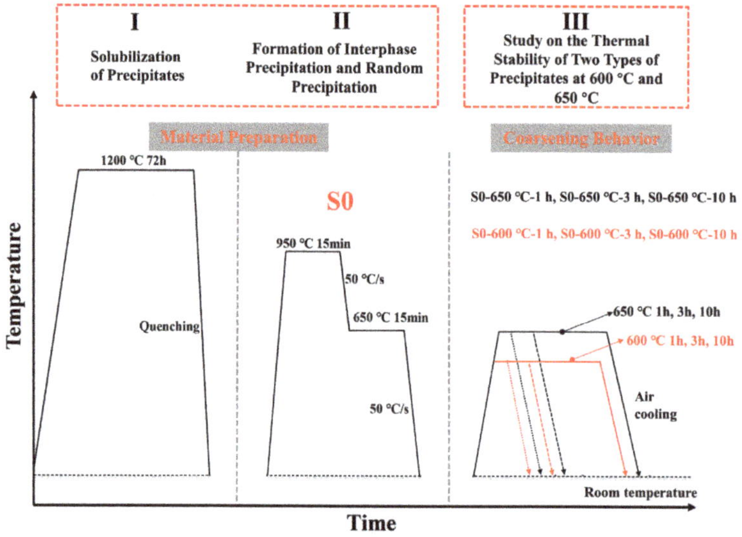

Figure 1. Schematic diagram illustrating the heat treatment process of the microalloyed steels.

3. Results

After being kept at a temperature of 1200 °C for a long period of time, the precipitated elements were basically all dissolved into the matrix. Then, after austenitizing at 950 °C and isothermal heat treatment at 650 °C, the microstructure shown in Figure 2 was obtained. The microstructure primarily consisted of ferrite, with a small amount of bainite present. This is consistent with the phase transformation behavior observed in the study [22]. The corresponding process is shown in stage II of Figure 1. During isothermal transformation at 650 °C, as described in the previous studies [23,24], due to the coordination of the moving speed of the front of the phase boundary and the enrichment of the precipitating elements on the slip planes of step planes, a large amount of interphase precipitation appeared during the early stage. However, it was difficult to observe interphase precipitation in the later stage, and a large amount of random precipitation occurred. After isothermal transformation at 650 °C for 15 min, the samples were subsequently held at 600 °C and 650 °C for 1 h, 3 h, and 10 h, respectively. The corresponding process is shown in stage III of Figure 1. At this stage, due to the higher diffusion rate and higher energy of grain boundaries, carbides tended to precipitate easily at the grain boundaries, resulting in a significant amount of precipitation at these locations [25]. However, this study primarily focuses on interphase and random precipitates within the grains and thus will not discuss this in further detail. The selection of these two temperatures was primarily based on the environmental temperatures encountered during the fire-resistant testing of the steel [26]. Through TEM observation and energy-dispersive X-ray spectroscopy (EDS) analysis, it can be seen that the precipitates in sample S0 were mainly interphase precipitation (as indicated by the white dashed lines in Figure 3a) and random precipitation (as indicated by the white circles in Figure 3a). The EDS analysis in Figure 3b confirms that the precipitates were primarily TiC, and previous research also verified that the carbide was the MC (M = Mo/Ti/V/Nb) carbide (Metallic Carbides) with an NaCl-type crystal structure, which will not be elaborated in this article [21–23,27,28].

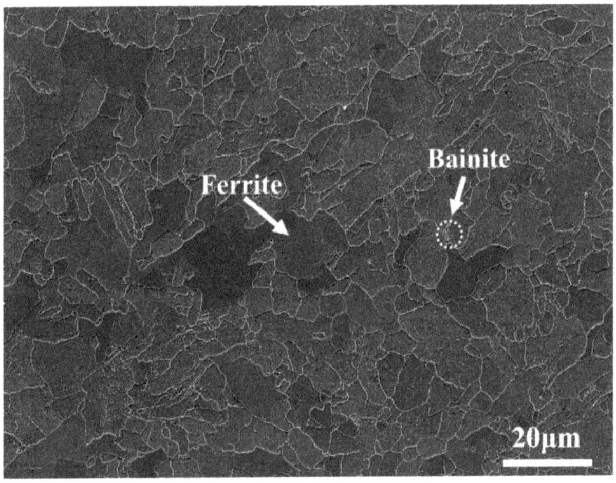

Figure 2. SEM micrographs of sample S0.

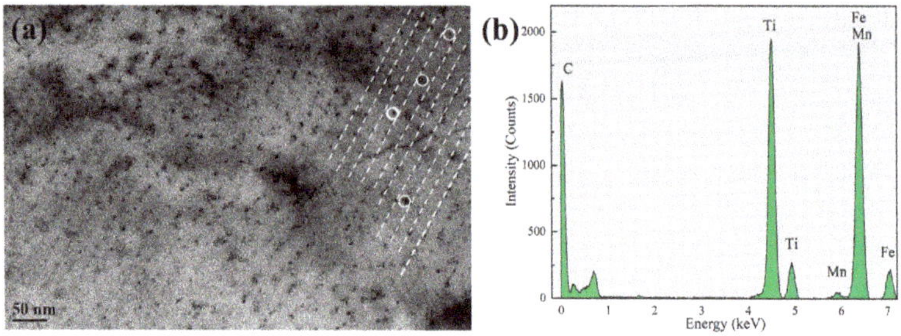

Figure 3. TEM micrographs (**a**) and EDS analysis (**b**) of the precipitates in sample S0.

In order to avoid the influence of thickness, SEM micrographs were used instead of TEM micrographs to observe the precipitates. Figures 4 and 5 present the SEM micrographs of precipitates at different holding times at 600 °C and 650 °C, along with corresponding images processed using the MATLAB program. We can observe that there are rows of interphase precipitation and random precipitation. The method of equivalent area circle was adopted to measure the size of the precipitates. A program written in MATLAB was utilized to count the number of pixel points of each independent precipitate (black spot) in the binarized image. The pixel count was then converted into the actual area using a calibration scale. Each precipitate was approximately regarded as a perfect circle, and its diameter was calculated based on the area, which was taken as the approximate diameter of the precipitate. Figures 4 and 5 clearly show that precipitate coarsening and growth occurred more slowly with partitioned precipitation compared to random precipitation. Additionally, higher temperatures accelerated this coarsening and growth process. After 10 h at 650 °C, the coarsening trend became particularly evident.

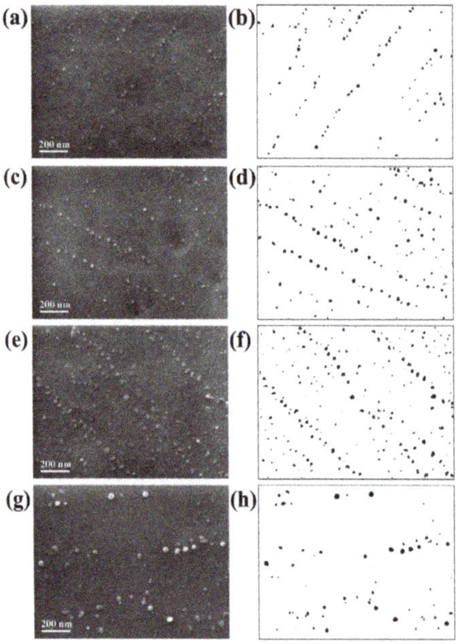

Figure 4. SEM micrographs (**a**,**c**,**e**,**g**) and binarized images (**b**,**d**,**f**,**h**) of precipitates with different isothermal holdings time at 600 °C in the microalloyed steels. (**a**,**b**) S0 (**c**,**d**) S0-600 °C-1 h, (**e**,**f**) S0-600 °C-3 h, (**g**,**h**) S0-600 °C-10 h.

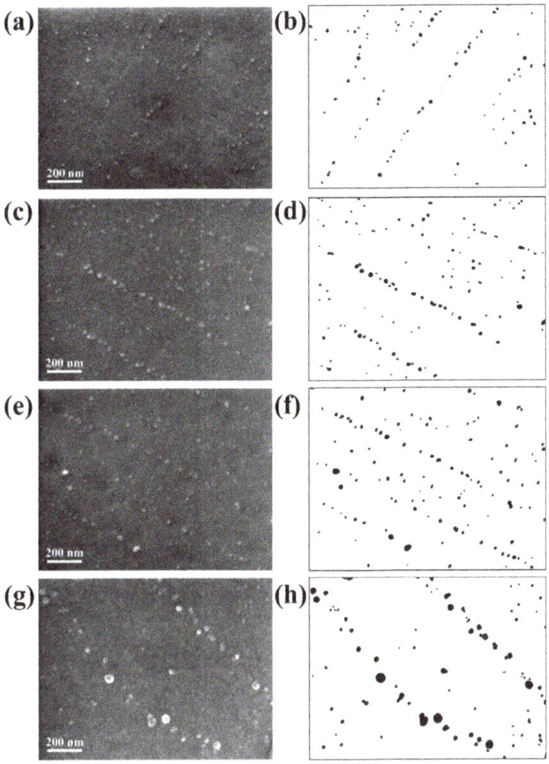

Figure 5. SEM micrographs (**a**,**c**,**e**,**g**) and binarized images (**b**,**d**,**f**,**h**) of precipitates with different isothermal holding times at 650 °C in the microalloyed steels. (**a**,**b**) S0, (**c**,**d**) S0-650 °C-1 h, (**e**,**f**) S0-650 °C-3 h, (**g**,**h**) S0-650 °C-10 h.

4. Discussion

4.1. Coarsening Kinetics

In order to investigate the coarsening behavior of different precipitates at various temperatures, it is essential to quantify the precipitate size at different holding times and temperatures. The precipitate size, represented by the equivalent circular diameter in this study, varies with the holding time, as shown in Figure 6. The error bars indicate the maximum and minimum values within the 99% confidence interval of the measured data.

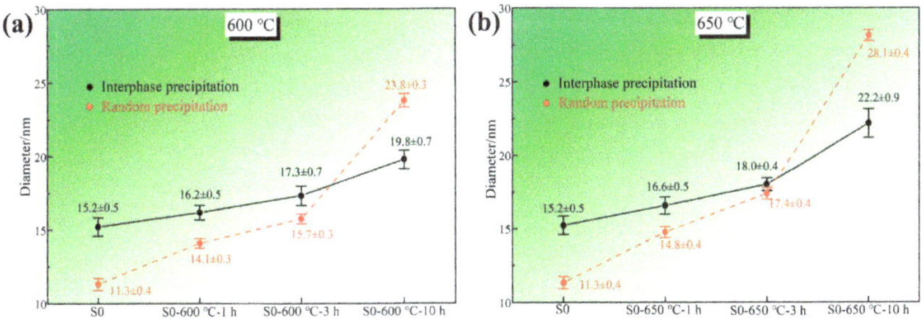

Figure 6. Variation of the average diameter of precipitates with different isothermal holding times at 600 °C (**a**) and 650 °C (**b**) in the microalloyed steels.

Interphase precipitation occurs at the phase boundary during the early stage of phase transformation when held at 650 °C for a period of time; its size is slightly bigger than random precipitation. According to statistical analysis, the average diameter of interphase precipitation of sample S0 was ~15.2 nm, while the random precipitation was ~11.3 nm. With the increase in holding time, the size of the interphase precipitates was gradually increased. The average diameter of the interphase precipitates was still only ~19.8 nm at 600 °C for 10 h, and increased by ~30%. The average diameter of the interphase precipitates was ~22.2 nm at 650 °C for 10 h, and increased by ~46%. This indicates that interphase precipitation is more likely to grow at 650 °C compared to 600 °C. In contrast, with the increase in holding time, the average size of random precipitates increased significantly at 600 °C, especially for 3 h to 10 h, and the average diameter increased by ~52%. During the initial three hours of the holding process, the precipitate size increased slowly. However, a noticeable acceleration in growth occurred after just one hour at 650 °C. From 1 and 10 h, the average diameter of the random precipitates grew by ~90%. Based on the above, interphase precipitation grew more slowly than random precipitation at both 600 °C and 650 °C, with no significant increase in average size over the entire 10-h period. In contrast, random precipitation at 600 °C exhibited a sharp increase in growth rate starting from 3 h. At 650 °C, this rapid growth phase began earlier, and the increase in growth rate was more pronounced.

The size of the precipitates shows a certain variation rule with time, and there is an equation proposed by some research on the variation of precipitate size with time [29].

$$\left(r_t^3 - r_0^3\right) = \frac{K}{RT} V^2 C D \gamma t \qquad (1)$$

where, r_t and r_0 represent the average particle radius at time 0 and t, respectively; D is the diffusion coefficient of the solute in the matrix; γ is the interfacial energy between the precipitate and the matrix; C is the solute concentration at equilibrium; V represents the molar volume of the precipitate; and K is a constant.

To fit the exponential relationship between the precipitation radius and time, Equation (1) is simplified, using N as the time exponent and K′ as the constant, to obtain Equation (2).

$$\left(r_t^N - r_0^N\right) = K' t \qquad (2)$$

Based on the fitting results of Equation (2) shown in Figure 7, the time exponent for random precipitates at 600 °C is N = 3 with a correlation of 0.982, and the time exponent for interphase precipitates at 600 °C is N = 6 with a correlation of 0.999. According to the statistical results in this paper, the average size of the random precipitates at 600 °C follows the 1/3 power law of t, while the average size of the interphase precipitates at 600 °C is close to the 1/6 power law of t. Similarly, based on the fitting results of Equation (2) shown in Figure 8, the time exponent for the random precipitates at 650 °C is N = 3 with a correlation of 0.977, and the time exponent for the interphase precipitates at 650 °C is N = 5 with a correlation of 0.993. According to the statistical results in this paper, the average size of the random precipitates at 600 °C or 650 °C follows the 1/3 power law of t, while the average size of the interphase precipitates at 600 °C is close to the 1/6 power law of t and at 650 °C is close to the 1/5 power law of t. According to relevant theories, the 1/3 power law of t applies to the later stage of second-phase particle precipitation in a supersaturated solid solution, where the coalescence process plays a dominant role. At this stage, the supersaturation of the matrix becomes very small, and the growth of the grains occurs primarily through the incorporation of smaller grains by larger ones [30–32]. The 1/4 power law of t applies to the precipitates on the grain boundary and the 1/5 power law of t applies to the precipitates on dislocation [30–32]. During the coarsening process, solute diffusion is governed by a characteristic distance related to the spatial distribution of the particles, which is influenced by the volume fraction of the precipitates [31,32]. Consequently, this affects the coarsening rate. It can be seen from the results that the exponent of the precipitate radius and time of random precipitates are consistent with those of supersaturated precipitates; the interphase precipitates are smaller than the precipitates on dislocation or the grain boundary; and the growth and coarsening rates of interphase precipitates are slower than that of random precipitates.

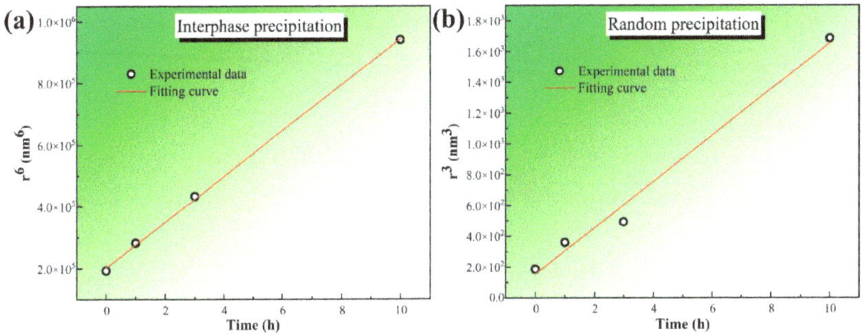

Figure 7. Plot of the mean radius of different precipitates against time at 600 °C: (**a**) is interphase precipitation; (**b**) is random precipitation.

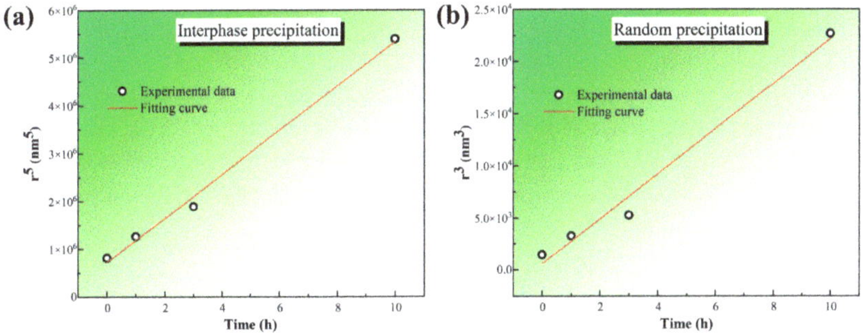

Figure 8. Plot of the mean radius of different precipitates against time at 650 °C: (**a**) is interphase precipitation; (**b**) is random precipitation.

4.2. The Effect of Precipitate Size Uniformity

As analyzed above, it can be seen that, whether at 600 °C or 650 °C, the growth and coarsening rates of interphase precipitation are significantly smaller than that of random precipitation. The same trend is observed during holding at both temperatures. Therefore, the following discussion will focus on analyzing the reasons behind the different growth and coarsening behaviors of the two types of precipitate at 600 °C.

First, an analysis will be conducted based on the density of the two types of precipitate. In view of the unique distribution, we adopted the linear density approach to determine the density of interphase precipitates. It can be seen from Figure 9 that the linear density of the interphase precipitates decreased from 3 h to 10 h at 600 °C, while the area density of the random precipitates decreased to a certain extent from 1 h to 3 h at 600 °C, and the decrease was more obvious after 3 h. Combined with the average diameter and density of precipitates, we can determine that the coarsening rate of interphase precipitates is significantly slower than that of random precipitates.

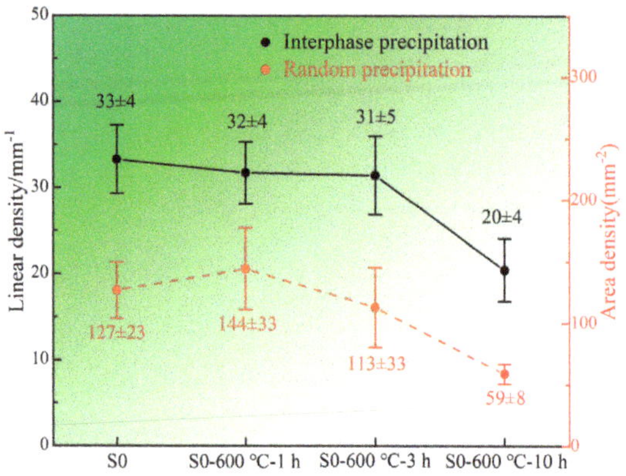

Figure 9. Variation of the average density of precipitates with different isothermal holding times at 600 °C in the microalloyed steels.

It is well known that interphase precipitation is formed on the slip planes of step planes of the phase boundary between ferrite and austenite. If the phase transformation temperature is constant, the size of the precipitates will be relatively uniform. If the phase transformation temperature is constant, the slip planes of step planes will disappear quickly, and the elements will not accumulate on it, which will lead to further growth of interphase precipitates. Thus, the size of the interphase precipitates is relatively uniform. However, due to the different formation times, the size difference of the random precipitates is obvious. From Figure 10, we can clearly identify that the full width at half maximum (FWHM) of the diameter distribution diagram of the interphase precipitates is significantly smaller than that of the random precipitates. The FWHM reflects the uniformity of precipitate sizes. A smaller FWHM indicates that the sizes of the precipitates within the statistical range are more consistent. Therefore, it can be concluded that the interphase precipitates have a more uniform size than the random precipitates.

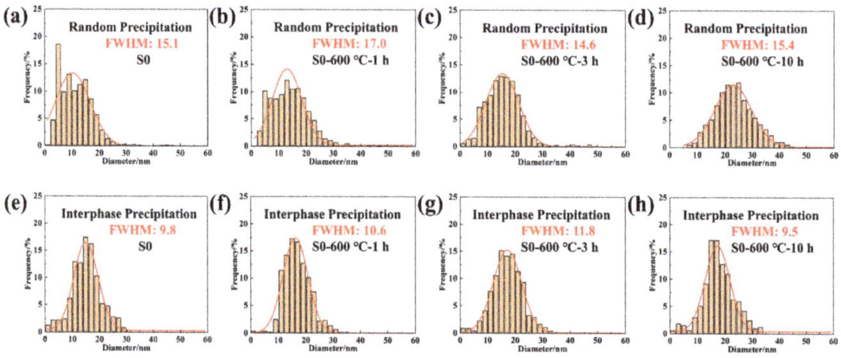

Figure 10. Distributions of precipitate diameter determined from the SEM images of the microalloyed steels held at 600 °C for various times: (**a**–**d**) are the statistical diagrams of random precipitation; (**e**–**h**) are the statistical diagrams of interphase precipitation (FWHM: the full width at half-maximum).

As is known to all, the coarsening of precipitates follows the Lifshitz–Slyozov–Wagner (LSW) theory [30,33]. LSW theoretical analysis is based on the Gibbs–Thompson equation, which is expressed by:

$$C_r = C_0 \exp\left[\frac{2\sigma V_m}{rRT}\left(\frac{1 - C_0}{C_p - C_0}\right)\right] \tag{3}$$

where C_r is the atomic fraction of solute in the matrix at the precipitate interfaces of radius r, σ is the specific interfacial energy of the matrix–precipitate boundary, V_m is the molar volume of the precipitate, C_0 is the equilibrium atomic fraction of Ti in the matrix, and C_p is the equilibrium atom fraction of Ti in the precipitates. The precipitate in this study is TiC, so the value of C_p is 1/2. However, the atomic fraction of Ti in the matrix is very low compared with that in the precipitates, and C_0 can be ignored. Thus, the exponential term can be simplified to $\frac{4\sigma V_m}{rRT}$. 4σ is ~1 J/m^2; the order of magnitude of V_m is ~10^{-5} m^3/mol, and the order of magnitude of RT is ~10^5 J/mol. Therefore, when the radius of the precipitate is greater than 0.1 nm, $\frac{4\sigma V_m}{rRT}$ is less than 1; the exponential function can be expanded in power series and the high-order term can be omitted such that

$$C_r - C_0 \approx C_0\left(\frac{4\sigma V_m}{rRT}\right) \quad (4)$$

From Equation (4), we can conclude that the smaller the size of precipitates, the higher the concentration of Ti in the precipitates. Figure 11 illustrates the differences in the size of the two types of precipitates and the atomic concentration profiles in their vicinity. The Ti atoms diffuse from the small-size precipitates to the large-size precipitates. This diffusion process destroys the concentration balance around the precipitates, resulting in the continuous dissolution and contraction of the small-size precipitates and the continuous growth of the large-size precipitates. However, the size distribution of the precipitates is uniform, and the concentration gradient does not occur easily. Accordingly, the coarsening rate of the interphase precipitates is slower than that of the random precipitates, and the thermal stability is higher.

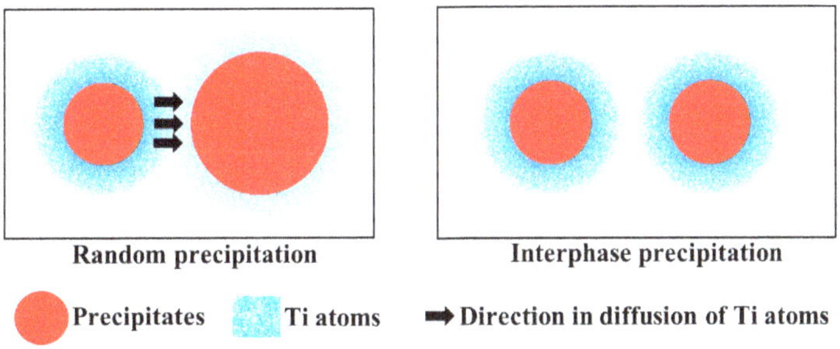

Figure 11. Schematic diagram of the diffusion of Ti atoms between interphase precipitation and random precipitation.

4.3. Properties and Evolution Behavior

Through observation and analysis, it can be seen that the coarsening growth rate of the interphase precipitates is slower than that of the random precipitates. The most important role of precipitates in steel is to strengthen through precipitation strengthening, and the level of the precipitation-strengthening contribution is mainly determined by the size and density of the precipitates. This paper estimates the precipitation-strengthening contribution $\Delta\tau_{orowan}$ through the Orowan–Ashby Equation (5) [34].

$$\Delta\tau_{orowan} = 0.26\left(\frac{Gb}{r}\right)f^{1/2}\ln\left(\frac{r}{b}\right) \quad (5)$$

where G is the shear modulus (80,300 MPa) [35], r is the average radius of the precipitates (shown in Figure 6a), b is the Burgers vector (0.248 nm), and f is the area percentage of the precipitates (the volume percentage of the precipitates is converted based on the linear density and area density in Figure 9). The results are shown in Table 2.

Table 2. The volume fraction of precipitates and precipitation-strengthening contribution.

Type of Precipitates	S0		S0-600 °C-1 h		S0-600 °C-3 h		S0-600 °C-10 h	
	f/%	$\Delta\tau_{orowan}$	f/%	$\Delta\tau_{orowan}$	f/%	$\Delta\tau_{orowan}$	f/%	$\Delta\tau_{orowan}$
RP	0.027	47	0.03	42.9	0.024	35.2	0.012	18.8
IP	0.035	43.6	0.033	41	0.032	38.5	0.027	28.1

As shown in Table 2, in sample S0, although the volume fraction is small due to the smaller size of the random precipitates, the precipitation-strengthening contribution of random precipitation is higher than that of interphase precipitation. As the tempering time at 600 °C increases, the precipitation-strengthening contribution of random precipitation decreases substantially. Compared to its value before tempering, the precipitation-strengthening contribution of interphase precipitation decreases from ~43.6 MPa to ~28.1 MPa after 10 h of tempering, a reduction of ~35.6%, while that of random precipitation decreases by ~60%. The calculation results indicate that interphase precipitation maintains a higher precipitation-strengthening contribution after prolonged tempering compared to random precipitation. It is precisely due to the superior thermal stability of interphase precipitation that the hardness of the experimental steel did not decrease sharply with the increase in holding time at 600 °C. The results of microhardness measurements are presented in Table 3.

Table 3. The Vickers hardness values of different samples.

Sample	S0	S0-600 °C-1 h	S0-600 °C-3 h	S0-600 °C-10 h
HV0.01	181.7 ± 4.5	175.6 ± 1.3	176.6 ± 1.9	174.0 ± 1.5

In conclusion, we can clearly understand the precipitation processes of interphase precipitation and random precipitation from Figure 12. Additionally, it also illustrates the growth and coarsening rules of interphase precipitation and random precipitation under different tempering times. When held at 950 °C, it is completely transformed into austenite, and then at 650 °C, the ferrite phase transformation occurs, forming interphase precipitation. Afterwards, random precipitation is formed within the ferrite. During the subsequent holding process at 600 °C or 650 °C, the interphase precipitation has superior thermal stability than random precipitation, and its growth and coarsening rates are slower.

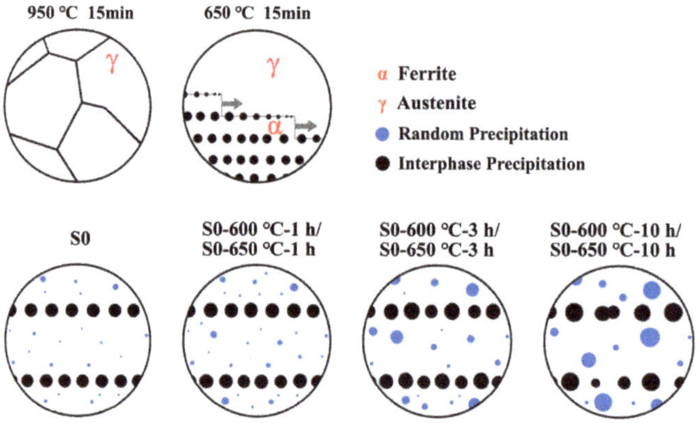

Figure 12. Schematic illustration of changes in the morphology and nature of precipitates during different processes.

5. Conclusions

In conclusion, this paper studies the thermal stability and coarsening behavior of interphase precipitation and random precipitation by analyzing the changes in size and density of different types of precipitate at elevated temperatures. The conclusions are summarized as follows.

(1) At an elevated temperature, the coarsening rates of interphase precipitates are significantly lower than random precipitates. As the temperature increases, the coarsening of interphase precipitates progresses at a slower rate.
(2) The coarsening behavior of random and interphase precipitates at 600 °C and 650 °C follows distinct power laws, with random precipitates adhering to a 1/3 power law and interphase precipitates following 1/6 and 1/5 power laws at 600 °C and 650 °C, respectively.
(3) Interphase precipitation forms on the slip planes at the phase boundary and exhibits a uniform size, with an FWHM of 9.8, whereas random precipitation shows a pronounced size distribution due to varying formation times, with an FWHM of 15.1. The uniform size of interphase precipitates makes it challenging to form a concentration gradient, leading to slower coarsening and enhanced thermal stability.

Author Contributions: Conceptualization, X.W. and X.X.; Methodology, X.W. and X.X.; Software, H.Z., X.S. and Z.S.; Validation, H.W., X.W. and X.X.; Formal analysis, J.C., Y.Y. and H.W.; Investigation, J.C., Y.Y., H.Z., X.S. and Z.S.; Data curation, J.C., Y.Y., H.Z., X.S. and Z.S.; Writing—original draft, J.C. and Y.Y.; Writing—review & editing, H.W., X.W. and X.X.; Visualization, H.Z., X.S. and Z.S.; Supervision, X.W. and X.X. All authors have read and agreed to the published version of the manuscript.

Funding: This research was funded by the Science and Technological Plan Project in the Inner Mongolia Autonomous Region, grant number 2023YFHH0033.

Data Availability Statement: The original contributions presented in this study are included in the article. Further inquiries can be directed to the corresponding authors.

Acknowledgments: The authors are also grateful for the important assistance provided by Jiajie Fan.

Conflicts of Interest: Author Zhendong Song was employed by Technical Center of Inner Mongolia Baotou Steel Union Co., Ltd. The remaining authors declare that the research was conducted in the absence of any commercial or financial relationships that could be construed as a potential conflict of interest.

·ferences

Xie, Z.; Song, Z.; Chen, K.; Jiang, M.; Tao, Y.; Wang, X.; Shang, C. Study of Nanometer-Sized Precipitation and Properties of Fire Resistant Hot-Rolled Steel. *Metals* **2019**, *9*, 1230. [CrossRef]
Kim, J.; Yu, G.; Kim, S.; Park, J.; Ahn, M.; Chung, J.-H.; Lee, C.-H.; Shin, C. Microstructural and Mechanical Characterization of Low-Alloy Fire- and Seismic-Resistant H-Section Steel. *Metals* **2024**, *14*, 374. [CrossRef]
Ferreira, P.P.; Carvalho, F.M.; Ariza-Echeverri, E.A.; Delfino, P.M.; Bauri, L.F.; Ferreira, A.M.; Braga, A.P.; Eleno, L.T.F.; Goldenstein, H.; Tschiptschin, A.P. Synergism between B and Nb Improves Fire Resistance in Microalloyed Steels. *Metals* **2023**, *13*, 84. [CrossRef]
Zhang, Z.; Niu, G.; Li, J.; Zhang, P.; Wu, H. Synergistic Effects of Ti and Mo on the Microstructure and Properties of Low-Cost Fire-Resistant and Weather-Resistant Construction Steel. *J. Mater. Eng. Perform.* **2023**, *32*, 3958–3967. [CrossRef]
Zhou, Y.; Liu, Y.; Zhou, X.; Liu, C.; Yu, J.; Huang, Y.; Li, H.; Li, W. Precipitation and Hot Deformation Behavior of Austenitic Heat-Resistant Steels: A Review. *J. Mater. Sci. Technol.* **2017**, *33*, 1448–1456. [CrossRef]
Wang, H.; Du, H.; Wei, Y.; Hou, L.; Liu, X.; Wei, H.; Liu, B.; Jia, J. Precipitation and Properties at Elevated Temperature in Austenitic Heat-Resistant Steels—A Review. *Steel Res. Int.* **2021**, *92*, 2000378. [CrossRef]
Zhang, Z.; Wang, Z.; Li, Z.; Sun, X. Microstructure Evolution and Precipitation Behavior in Nb and Nb-Mo Microalloyed Fire-Resistant Steels. *Metals* **2023**, *13*, 112. [CrossRef]

8. Wang, H.; Yan, W.; Van Zwaag, S.; Shi, Q.; Wang, W.; Yang, K.; Shan, Y. On the 650 °C Thermostability of 9–12Cr Heat Resista Steels Containing Different Precipitates. *Acta Mater.* **2017**, *134*, 143–154. [CrossRef]
9. Dudko, V.; Belyakov, A.; Molodov, D.; Kaibyshev, R. Microstructure Evolution and Pinning of Boundaries by Precipitates ir 9 Pct Cr Heat Resistant Steel During Creep. *Metall. Mater. Trans. A* **2013**, *44*, 162–172. [CrossRef]
10. Jayanth, C.S.; Nash, P. Factors Affecting Particle-Coarsening Kinetics and Size Distribution. *J. Mater. Sci.* **1989**, *24*, 3041–30 [CrossRef]
11. Martin, J.W. *Precipitation Hardening: Theory and Applications;* Butterworth-Heinemann: Oxford, UK, 2012.
12. Guo, Z.; Sha, W. Quantification of Precipitation Hardening and Evolution of Precipitates. *Mater. Trans.* **2002**, *43*, 1273–12 [CrossRef]
13. Yin, Y.F.; Faulkner, R.G. Modelling the Effects of Alloying Elements on Precipitation in Ferritic Steels. *Mater. Sci. Eng. A* **2003**, 3 92–102. [CrossRef]
14. Kesternich, W.; Meertens, D. Microstructural Evolution of a Titanium-Stabilized 15Cr-15Ni Steel. *Acta Metall.* **1986**, *34*, 1071–10 [CrossRef]
15. Miyata, K.; Sawaragi, Y.; Okada, H.; Masuyama, F.; Yokoyama, T.; Komai, N. Microstructural Evolution of a 12Cr-2W-Cu-V-N Steel during Three-Year Service Exposure. *ISIJ Int.* **2000**, *40*, 1156–1163. [CrossRef]
16. Funakawa, Y. Interphase Precipitation and Application to Practical Steels. *Mater. Trans.* **2019**, *60*, 2086–2095. [CrossRef]
17. Zhang, Y.; Shinbo, K.; Ohmura, T.; Suzuki, T.; Tsuzaki, K.; Miyamoto, G.; Furuhara, T. Randomization of Ferrite/Austen Orientation Relationship and Resultant Hardness Increment by Nitrogen Addition in Vanadium-Microalloyed Low Carbon Ste Strengthened by Interphase Precipitation. *ISIJ Int.* **2018**, *58*, 542–550. [CrossRef]
18. Wang, J.; Hodgson, P.D.; Bikmukhametov, I.; Miller, M.K.; Timokhina, I. Effects of Hot-Deformation on Grain Bounda Precipitation and Segregation in Ti-Mo Microalloyed Steels. *Mater. Des.* **2018**, *141*, 48–56. [CrossRef]
19. Funakawa, Y.; Shiozaki, T.; Tomita, K.; Yamamoto, T.; Maeda, E. Development of High Strength Hot-Rolled Sheet Steel Consisti of Ferrite and Nanometer-Sized Carbides. *ISIJ Int.* **2004**, *44*, 1945–1951. [CrossRef]
20. Yang, Y.-L.; Lin, Q.-X.; Chen, C.-Y.; Jen, Y.-M.; Wang, S.-H.; Chen, H.-R.; Tsai, B.-S.; Liu, H.-C. Insight on Creep Behavior f Arrayed Interphase Nano-Precipitation in Ti-Microalloyed HSLA Steel. *Mater. Chem. Phys.* **2020**, *255*, 123553. [CrossRef]
21. Cong, J.; Li, J.; Fan, J.; Liu, P.; Misra, R.D.K.; Shang, C.; Wang, X. The Impact of Interphase Precipitation on the Mechanic Behavior of Fire-Resistant Steels at an Elevated Temperature. *Materials* **2020**, *13*, 4294. [CrossRef] [PubMed]
22. Krbata, M.; Krizan, D.; Eckert, M.; Kaar, S.; Dubec, A.; Ciger, R. Austenite Decomposition of a Lean Medium Mn Steel Suitable f Quenching and Partitioning Process: Comparison of CCT and DCCT Diagram and Their Microstructural Changes. *Materials* **20** *15*, 1753. [CrossRef]
23. Chih-Yuan, C.; Shih-Fan, C.; Chien-Chon, C.; Jer-Ren, Y. Control of Precipitation Morphology in the Novel HSLA Steel. *Mater. S Eng. A* **2015**, *634*, 123–133. [CrossRef]
24. Chen, C.-Y.; Yang, J.-R.; Chen, C.-C.; Chen, S.-F. Microstructural Characterization and Strengthening Behavior of Nanomet Sized Carbides in Ti–Mo Microalloyed Steels during Continuous Cooling Process. *Mater. Charact.* **2016**, *114*, 18–29. [CrossRef
25. Brytan, Z.; Król, M.; Benedyk, M.; Pakieła, W.; Tański, T.; Dagnaw, M.J.; Snopiński, P.; Pagáč, M.; Czech, A. Microstructural ar Mechanical Properties of Novel Co-Free Maraging Steel M789 Prepared by Additive Manufacturing. *Materials* **2022**, *15*, 17 [CrossRef]
26. E28 Committee ASTM International. *Test Methods for Elevated Temperature Tension Tests of Metallic Materials;* ASTM Internation West Conshohocken, PA, USA, 2009.
27. Jang, J.H.; Lee, C.-H.; Heo, Y.-U.; Suh, D.-W. Stability of (Ti, M)C (M = Nb, V, Mo and W) Carbide in Steels Using First-Principl Calculations. *Acta Mater.* **2012**, *60*, 208–217. [CrossRef]
28. Cong, J.; Li, J.; Fan, J.; Misra, R.D.K.; Xu, X.; Wang, X. Effect of Austenitic State before Ferrite Transformation on the Mechanic Behavior at an Elevated Temperature for Seismic-Resistant and Fire-Resistant Constructional Steel. *J. Mater. Res. Technol.* **2021**, 1220–1229. [CrossRef]
29. Honeycombe, R.W.K.; Mehl, R.F. Transformation from Austenite in Alloy Steels. *Met. Trans. A* **1976**, *7*, 915–936. [CrossRef]
30. Lifshitz, I.M.; Slyozov, V.V. The Kinetics of Precipitation from Supersaturated Solid Solutions. *J. Phys. Chem. Solids* **1961**, *19*, 35– [CrossRef]
31. Dunlop, G.L.; Honeycombe, R.K.W. The Role of Dislocations in the Coarsening of Carbide Particles Dispersed in Ferrite. *Phil Mag.* **1975**, *32*, 61–72. [CrossRef]
32. Ardell, A.J. On the Coarsening of Grain Boundary Precipitates. *Acta Metall.* **1972**, *20*, 601–609. [CrossRef]
33. Deschamps, A.; Hutchinson, C.R. Precipitation Kinetics in Metallic Alloys: Experiments and Modeling. *Acta Mater.* **2021**, 2 117338. [CrossRef]

Gladman, T. Precipitation Hardening in Metals. *Mater. Sci. Technol.* **1999**, *15*, 30–36. [CrossRef]
Lu, S.-L.; Yang, S.-C.; Zhu, K.; Chen, Y.-S.; Cairney, J.M.; Lin, C.-M.; Yen, H.-W. Strategy of Complex Carbides in Microalloyed 4Mn Steels Processed by Quenching and Austenite Reversion. *Mater. Des.* **2023**, *230*, 111951. [CrossRef]

claimer/Publisher's Note: The statements, opinions and data contained in all publications are solely those of the individual ıor(s) and contributor(s) and not of MDPI and/or the editor(s). MDPI and/or the editor(s) disclaim responsibility for any injury to ple or property resulting from any ideas, methods, instructions or products referred to in the content.

metals

Article

An Alternative Method of Investigating the Thermal Stability of Shoe-Braked Railway Wheel Steels Based on Strain Hardening Analysis

Giuliano Angella [1], Lorenzo Ghidini [2] and Michela Faccoli [2,*]

[1] CNR-ICMATE Research Institute, Milan Unit, Via R. Cozzi 53, 20125 Milan, Italy; giuliano.angella@cnr.it
[2] Department of Mechanical and Industrial Engineering, University of Brescia, Via Branze 38, 25123 Brescia, Italy; lorenzo.ghidini@unibs.it
* Correspondence: michela.faccoli@unibs.it

Abstract: During service, shoe-braked railway wheel steels are often subjected to a severe thermal cycle. Therefore, understanding the evolution of the microstructure and the resulting changes in mechanical properties during service is fundamental in the choice of steel. In previous research, the effects of the thermal loading on the microstructure and mechanical properties of five different steels for railway wheels (ER7, HYPERLOS®, Class B, SANDLOS® and Class C) were investigated by hardness, tensile and toughness tests, in the as-supplied condition and after different heat treatments designed to replicate the modification of the microstructure due to braking. In this paper, the tensile work hardening behavior was studied by interpolating the tensile flow curves with the constitutive equation related to the dislocation density proposed by Voce, which correlates the Voce equation parameters with the microstructural features of metallic materials. The work hardening analysis revealed that there is a good correlation between the Voce parameters and the microstructure of the five steels in as-supplied condition and after heat treatments. An interesting correlation was found between Voce parameters and apparent fracture toughness. After heat treatments at 700 °C and 750 °C the properties of the steels decreased, which was consistent with the evolution of the microstructure. However, after exposure at 970 °C with subsequent cooling in air, Class C steel appears to have a microstructure similar to the original microstructure, with tensile and toughness properties very similar to the as-supplied condition, demonstrating better microstructural stability compared to steels ER7, HYPERLOS®, Class B and SANDLOS®.

Keywords: railway wheel steel; tread braking; thermal stability; tensile testing; strain hardening analysis

1. Introduction

During operation, railway wheels experience microstructural changes in their surface layers, which have a significant impact on their performance and longevity. The material integrity is particularly challenged by the thermal cycle that wheel treads undergo during block braking, which alternates between heating from friction with the brake shoe and cooling from rail contact and is extremely accentuated by the formation of hot spots that have even higher temperatures [1–3]. This cyclic loading can cause significant tensile stresses [4,5], microstructural alterations such as pearlite spheroidization [6–9] or the production of white etching layers [10–12], roughness caused by wear processes on the tread leading to undesirable tread profiles [13–19] or cracks [20–22], possibly degrading until the failure of the component. Therefore, choosing and optimizing the right steel compositions for the production of railway wheels requires a deep understanding of the evolution of microstructure and its subsequent impacts on mechanical behavior.

Mechanical properties, like tensile, fatigue, thermal-fatigue, wear resistance and fracture toughness are paramount information to design the safe applications of metallic materials. Knowing the correlations between mechanical properties, microstructure and production route makes it possible to establish manufacturing parameters for a cost-effective high-quality industrial production. However, the microstructure characterization can be a time- and energy-effective procedure and quite often the gathered information is qualitative and can barely achieve some quantitative relationship with mechanical properties, unless one uses expensive and unpractical techniques for industrial purposes, like electron microscopy or micro-tomography, for instance. These microanalytical techniques have high resolution power, so they need a lot of statistics and the identification of the relevant aspects of their results is not always straightforward. On the other side, the acquirement of mechanical properties involves high material volumes, like the gauge of tensile specimens, for instance, so they represent reliable integral information according to the international standards.

Tensile testing gives the fundamental information for designing the applications of materials, like railway wheels' steel, and is the simplest test. So, a procedure based on tensile properties for assessing the quality of steels and metallic materials should be desirable and give solid relationships between the relevant microstructure features and the tensile information. At the same time, a relationship between this assessment and other mechanical properties, like fatigue resistance or fracture toughness, for instance, should be hopefully established, being very useful for industrial production valuation and designing potential applications of steels. Indeed, besides the tensile mechanical properties, like Yield Stress (YS), Ultimate Tensile Stress (UTS) and ductility (A5), other information can be gathered by the analysis of the tensile strain hardening, analysis that is based on the determination of the strain hardening rate, that is, $d\sigma/d\varepsilon_p$ where σ is the true stress and ε_p is the true plastic strain. Some constitutive equations that fit the tensile flow curves have physical base, like the Voce equation, for instance, and their fitting procedure is based on strain hardening analysis [23]. Furthermore, the best fitting equation parameters relate to the relevant microstructure features of the metallic materials, which could give rise to a quality-assessment procedure that is successfully used in foundry alloys [24,25]. In the present investigation, the quality assessment procedure based on strain hardening and the Voce equation was used to analyze and to relate the results of microstructure (grain size and phases) taken through conventional metallography, experimental hardness, tensile and toughness tests. These tests, conducted by Faccoli et al. [26,27], involved five railway wheel steels in the as-supplied condition and three different heat treatments designed to replicate the modification of the microstructure due to braking. The large amount of microstructure and mechanical data reported in [26,27] is not here reported to avoid redundance of information, while focus on the results of the new assessment procedure of material quality applied to the railway steels is the subject of this work. The novel procedure based on strain hardening analysis and the Voce constitutive equation has been here used for the first time on wrought steels rather than foundry materials for which it was designed. Only the summaries of the relevant microstructure and mechanical properties found in [26,27] are reported in the present work, for clarity's sake. The correlation between microstructure features of the steels and the best fitting Voce parameters is reported, as well as the relation with Brinell hardness measurements (HB) and apparent fracture toughness (K_Q) in controlled conditions. The quality-assessment procedure based on strain hardening and the Voce equation supported the evidence that investigated steels have a different microstructure and mechanical stability when undergoing severe thermal cycles.

2. Materials and Methods

A summary of the experimental tests previously conducted (see [26,27] for more details), as well as the analyses performed to derive the parameters of interest, is presented below.

2.1. Steels and Heat Treatments

ER7, HYPERLOS®, Class B, SANDLOS® and Class C forged railway wheel steels, supplied by Lucchini RS (Lovere, Bergamo, Italy), were investigated in the as-supplied condition and after heat treatment. The as-supplied condition of the wheel refers to the industrial process involving rim chilling followed by tempering heat treatments (for details, see [26]). The heat treatment involves heating to a pre-selected temperature at a rate of 40 °C per hour, maintaining this temperature for 45 min, and then cooling to room temperature in still air. The maintaining temperatures for the heat treatments—700 °C, 750 °C and 970 °C—were determined based on simulation results [26] and represent the regions of the iron–iron carbide phase diagram below the eutectoid temperature A_1, between critical temperatures A_1 and A_3, and above A_3, respectively which are reached in the wheel rim during drag braking. The chemical compositions of the steels are reported in Table 1.

Table 1. Chemical composition wt.% of the investigated steels.

Steel	C	Si	Mn	V	S	P
ER7	0.49	0.34	0.75	0.008	0.002	0.008
HYPERLOS®	0.51	0.38	0.78	0.038	0.002	0.015
Class B	0.65	0.26	0.63	0.005	0.001	0.012
SANDLOS®	0.63	0.88	0.84	0.008	0.001	0.009
Class C	0.74	0.34	0.80	0.004	0.001	0.006

2.2. Hardness Testing

In [26,27], Brinell hardness measurements (HB) were carried out on the radial section of the rim at a depth of 35 mm under the running surface (corresponding to the limit of wear or last turning diameter) complying with the EN 13262 Standard [28]. The tests were conducted using a 5 mm diameter ball, under a load of 7355 N, and with a dwell time of 15 s.

2.3. Tensile Testing and Tensile Data Treatment

Tensile tests were conducted in accordance with the EN 13262 Standard [28] using round tensile specimens. These specimens were extracted from a depth of 15 mm below the wheel running surface, with their axes aligned parallel to the tread. The strain gauge was 50 mm with a 10 mm wide diameter. The engineering stress S and engineering strain e of the tensile flow curves were transformed into true stress $\sigma = S\cdot(1 + \varepsilon)$ and true strain $\varepsilon = \ln(1 + e)$, respectively, by using engineering data up to UTS, as beyond UTS localized deformation at necking occurred, and the true stress and strain are not representative of the strain gauge plastic deformation. Finally, the true plastic strain ε_p was calculated by subtracting the elastic strains from the strains, according to the equation $\varepsilon_p = \varepsilon - \sigma/E$, where E is the experimental Young modulus. So, the plastic flow curves were fitted through the Voce equation that is defined as follows:

$$\sigma = \sigma_V + (\sigma_o - \sigma_V)\cdot e^{-\frac{\varepsilon_p}{\varepsilon_c}} \quad (1)$$

The Voce equation is an exponential decay equation, where σ_V is the saturation stress and ε_c is the critical strain that defines the rate with which the σ_V is achieved. σ_o is the back extrapolated stress at zero plastic strain, $\varepsilon_p = 0$. In order to fit Equation (1) to the experimental flow curves, that is, find the Voce parameters σ_V, σ_o and ε_c, the strain hardening analysis was investigated. In fact, the differential form of Equation (1) is given by:

$$\frac{d\sigma}{d\varepsilon_p} = \Theta_o - \frac{\sigma}{\varepsilon_c} \quad (2)$$

Θ_0 is a constant, equal to σ_V/ε_c, while ε_c has the usual meaning. So, if Equation (1) is the proper constitutive equation describing the experimental flow curve, through plotting the experimental strain hardening rate ($\frac{d\sigma}{d\varepsilon_p}$) vs. the true stress σ, a linear data region should be found, and through fitting this linear region with Equation (2), the parameters Θ_0 and ε_c are found according to the procedure reported in [24,25]. Finally, $\sigma_V = \Theta_0 \cdot \varepsilon_c$, while σ_0 is found through minimizing the difference between Equation (1) and the experimental flow curve.

According to a wide body of literature [23], the Voce parameters Θ_0 and ε_c have physical meaning. Θ_0 is athermal by nature, and describes the dislocation multiplication that is related to the mean distance λ between obstacles to dislocation motion; λ is related to the coarseness of the microstructure: the finer the microstructure (short λ), the bigger the Θ_0. So, bainitic structure has a smaller λ and bigger Θ_0 than a pearlitic structure, for instance, while finer pearlite grain size is described by a bigger Θ_0. The final value of Θ_0 depends on the statistical contribution of all the obstacles to dislocation motion according to a qualitative equation like

$$\Theta_0 \propto \alpha_1 \cdot \frac{V_1}{\lambda_1} + \alpha_2 \cdot \frac{V_2}{\lambda_2} + \alpha_3 \cdot \frac{V_3}{\lambda_3} + \ldots \quad (3)$$

where V_i and λ_i represent the volume fractions and the mean distance of the i-th obstacle of the microstructure, respectively, and α_i are multiplicative factors. The contribution from every single obstacle (α_i) cannot be easily found, and Equation (3) represents a qualitative relationship: different microstructures could have similar Θ_0, and so Θ_0 is a parameter describing the effective microstructure. $1/\varepsilon_c$ is thermal by nature and is related to a set of micro mechanisms that soften the deformed materials via reducing the dislocation density, like dislocation annihilation and formation of a low-energy dislocation structure, which depends on the dislocation motion: different crystallographic matrices and different chemical compositions have different values of $1/\varepsilon_c$.

In foundry products, it was found that by plotting the Voce parameters $1/\varepsilon_c$ vs. Θ_0 of a set of significant tensile tests of the same material, they lie alongside a single line. The distribution of these Voce parameters on a line depends on the natural scattering of microstructure features, that is, the scattering of the characteristic lengths λ_i, while the dislocation activity of dislocation multiplication and dynamic recovery is the same for all tensile flow curves as the matrix is unchanged. The plot $1/\varepsilon_c$ vs. Θ_0 is called Matrix Assessment Diagram (MAD), and on it metallic materials with different chemical compositions and/or different crystallographic structures lie on distinct lines that are like fingerprints [24]. For sound materials like wrought steels the intercept is positive, while in defective materials like some foundry products the best fitting lines of the Voce parameters in MAD have negative intercepts. The more defective the material, the more negative the intercept is, and this consistent behavior is called Defects-Driven Plasticity [29,30].

2.4. Toughness Testing and Metallographic Observations

The EN 13262 Standard mandates a minimum toughness for ER7 tread-braked wheels to prevent any type of undesired brittle fracture and ensure safety in service. The current standards do not specify requirements for fracture toughness for wheels made from the other steels investigated. The toughness tests were performed in accordance with the ASTM E399 Standard [31]. Compact Tension specimens with a thickness of 30 mm (CT30) and a width of 60 mm were machined from the wheel rim. Six samples were extracted from each wheel in the as-supplied condition, spaced every 60 degrees along the tangential direction (for more details, see [26,27]). For each heat-treated condition, three samples were tested, each extracted from positions 120 degrees apart. At the conclusion of the test, the specimen was sectioned along the mid-plane orthogonal to the fracture surface, parallel to the wheel tread. The resulting section was prepared using standard metallographic techniques and was chemically etched with 2% Nital to reveal the microstructure. Microstructural observations and pearlite grain size measurements, complying with ASTM E112 [32], were conducted using an optical microscope. All the results in [26,27] are summarized in Table 2 for the steels in as-supplied condition, while for the steels after heat treatments the results in [26,27] are summarized in Table 3.

Table 2. Microstructure, mechanical test results (adapted from Refs. [26,27]) and best fitting Voce parameters of the investigated steels in the as-supplied condition.

Steel	Microstructure [1]	Pearlite Grain Size G	HB	YS (MPa)	UTS (MPa)	A5 (%)	Θ_o (MPa)	$1/e_c$	K_Q ($MPa \cdot m^{1/2}$)
ER7	P + F + B	8.0	258 ± 4	610	911	14	35,273	32.1	89 ± 6
HYPERLOS®	P + F + B	8.5	256	568	885	28	23,549	21.4	102 ± 14
Class B	P + F	7.5	290 ± 4	659	1035	15	47,820	38.5	58 ± 7
SANDLOS®	P + F + B	8.5	300 ± 4	687	1148	14	55,910	40.7	49 ± 3
Class C	P + F + B	8.0	340 ± 4	730	1140	15	55,046	37.8	48 ± 4

[1] F = ferrite; P = pearlite; B = bainite (traces).

Table 3. Microstructure, mechanical test results (adapted from Refs. [26,27]) and the best fitting Voce parameters of the investigated steels after the heat treatments.

Steel	T (°C)	Microstructure [1]	Pearlit Grain Size G	HB	YS (MPa)	UTS (MPa)	A5 (%)	Θ_o (MPa)	$1/e_c$	K_Q ($MPa \cdot m^{1/2}$)
ER7	700	P + Pg + F	7.5	201 ± 1	451	693	24	17,331	19.5	75 ± 1
	750	P + Pg + F + B	7.5	190 ± 2	424	692	24	16,557	18.4	72 ± 3
	970	P + F	5.0 to 7.5	229 ± 3	467	837	17	31,354	30.9	66 ± 11
HYPERLOS®	700	Pg + F + B	8.5	198	444	674	26	14,642	17.4	103 ± 4
	750	P + Pg + F + B	11.5	202	444	712	19	15,041	16.6	99 ± 1
	970	P + F	5.0	226	493	856	19	25,136	24.3	63 ± 4
Class B	700	P + Pg + F	7.0	209 ± 3	453	756	22	27,226	29.5	64 ± 4
	750	P + Pg + F	7.0	202 ± 2	429	726	24	24,274	27.1	65 ± 12
	970	P + F	6.5	231 ± 8	437	866	15	36,135	34.1	60 ± 6
SANDLOS®	700	P + Pg + F + B	7.5	242 ± 4	501	869	23	27,015	23.5	54 ± 5
	750	P + Pg + F + B	7.5	230 ± 4	490	814	19	26,721	26.8	59 ± 5
	970	P + F + B + M	6.5	287 ± 15	620	1049	18	41,116	32.6	50 ± 6
Class C	700	P + Pg + F + B	7.5	246 ± 3	516	911	19	32,637	29.4	47 ± 2
	750	P + Pg + F	7.5	261 ± 2	474	858	21	28,211	27.4	48 ± 2
	970	P + F + B + M	7.0	326 ± 9	674	1146	13	57,079	41.8	49 ± 1

[1] F = ferrite; P = pearlite; Pg = globular pearlite; B = bainite (traces); M = martensite (traces).

3. Results

3.1. Steels in the As-Supplied Condition

The microstructural observation results, HB measurements, tensile properties, fracture toughness taken from [26,27] and the calculated best fitting Voce parameters of the steels in the as-supplied condition are reported in Table 2.

In Figure 1, the best fitting Voce parameters of the steels in the as-supplied condition are reported as $1/\varepsilon_c$ vs. Θ_o in the MAD. The data lay close to a single line, suggesting that the different chemical compositions and production routes had significant effects on the different microstructural constituents of the steels, observed in [26,27] and summarized in Table 2, like the presence of ferrite (F), lamellar pearlite (P), bainite (B) and pearlite grain size (G), but there were no significant effects on the dislocation activity, i.e., on the soft ferrite where dislocation multiplication and dynamic recovery occurred, even if the chemical compositions of the steels were different. Besides Class B which had only ferrite and pearlite, the other steels also had traces of bainite and similar pearlite grain size, as reported in Table 2. The main alloying elements of the five investigated steels were C, Mn and Si; in addition, some of them were micro-alloyed with V. While C and Mn have the main effects on volume fractions and fineness of pearlite and bainite, Si partitions mainly in ferrite, strengthening it through the solid solution mechanism. In HYPERLOS®, V is added since it has been reported [32,33] that V forms carbides, strengthening the ferritic matrix, but preserving ductility and toughness. So, the different chemical compositions of the steels did not affect significantly the ferrite where dislocations move, multiply and dynamically recover, resulting in Voce parameters lying alongside the same linearity in MAD.

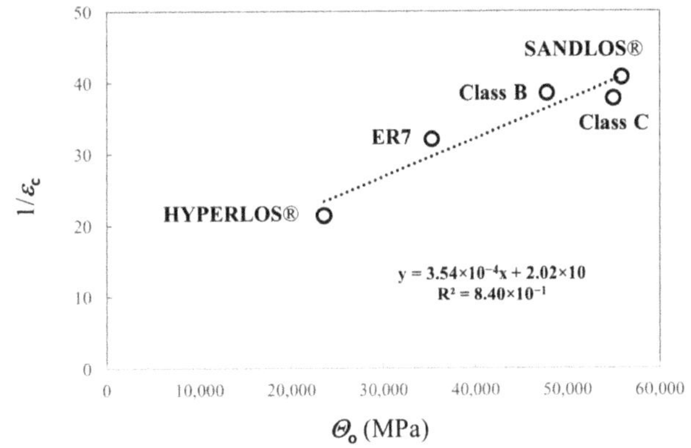

Figure 1. Matrix Assessment Diagram (MAD) of the five steels tensile-tested in the as-supplied condition.

In Figure 2, selected mechanical properties were plotted vs. the best fitting Voce parameter Θ_0. Linear relationships were found, proving that the Voce parameter Θ_0 was really related to the coarseness of the microstructure. In fact, the finer the microstructure, the higher the hardness and the YS, and so the inverse linear trend between Θ_0 and the values of HB and YS confirmed the validity of Equation (3).

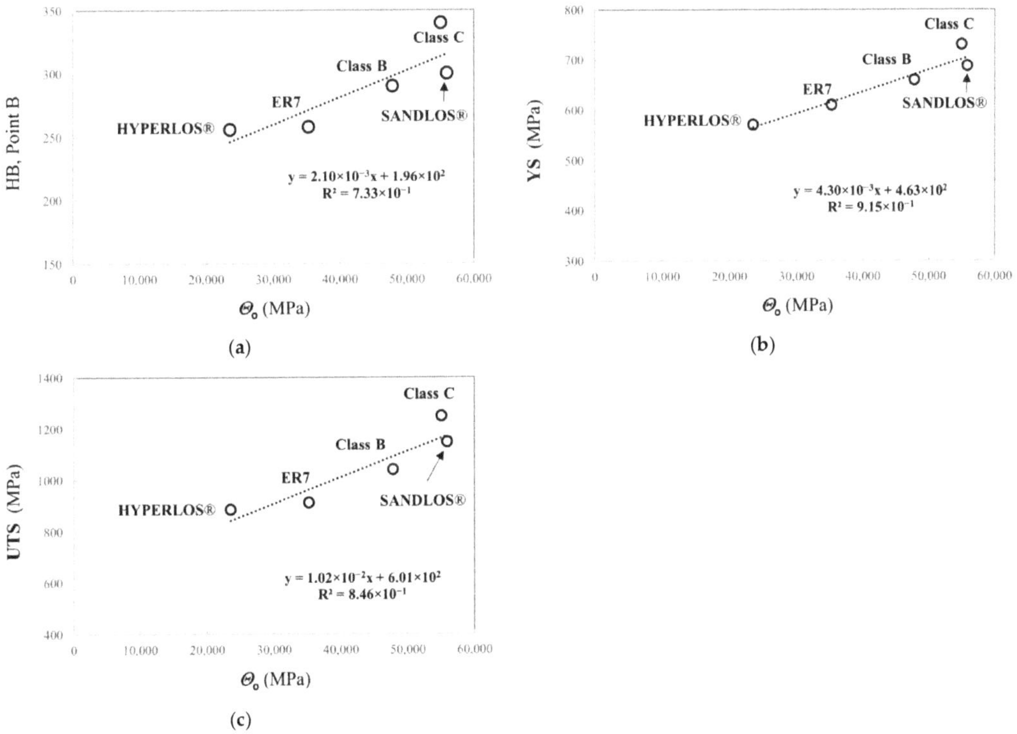

Figure 2. Comparison between selected mechanical properties and the best fitting Voce parameter Θ_0 of the steels in the as-supplied condition: (**a**) Brinell Hardness (HB) vs. Θ_0; (**b**) yield stress (YS) vs. Θ_0; (**c**) ultimate tensile stress (UTS) vs. Θ_0.

In Figure 3, the apparent fracture toughness K_Q values are reported vs. the Voce parameter Θ_o of the steels in the as-supplied condition. A good linearity was found between K_Q and Θ_o for the steels. The apparent fracture toughness K_Q was reported vs. HB hardness and YS in Figure 4a–c, respectively, where for K_Q vs. HB, K_Q vs. YS and K_Q vs. UTS good linear trends were also found.

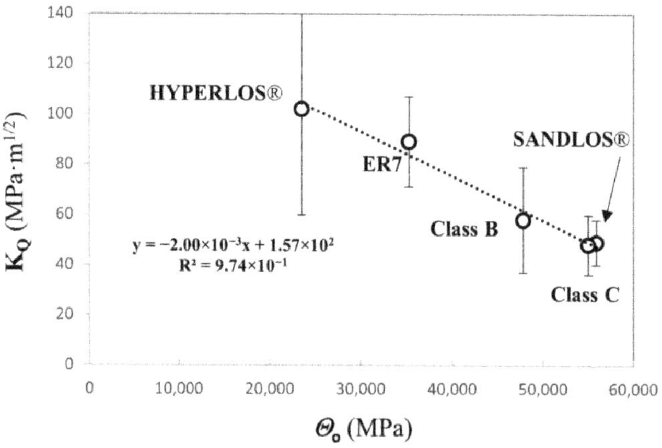

Figure 3. Apparent fracture toughness K_Q vs. best fitting Voce parameter Θ_o of the steels in as-supplied condition.

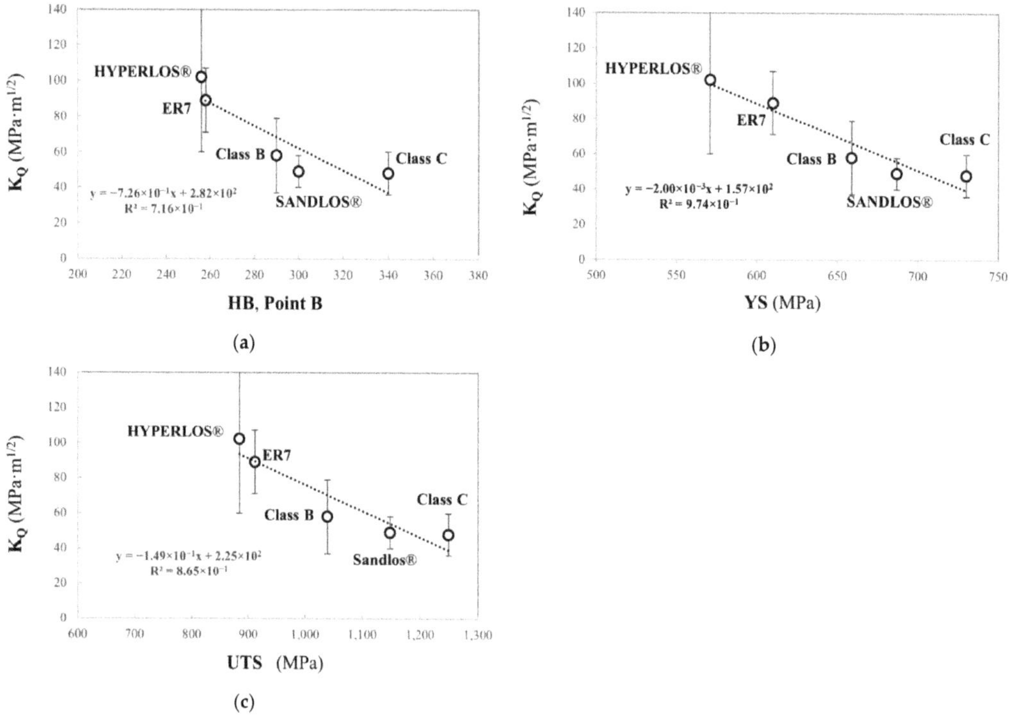

Figure 4. Apparent fracture toughness K_Q vs. selected mechanical properties of the investigated steels in the as-supplied condition: (**a**) K_Q vs. HB; (**b**) K_Q vs. YS; (**c**) K_Q vs. UTS.

3.2. *Steels after Heat Treatments*

In Table 3, the calculated best fitting Voce parameters of the five steels after heat treatments are reported, with the microstructure, the pearlite grain size, HB measurements, tensile properties and fracture toughness taken from [26,27]. After heat treatments at 700 and 750 °C, the microstructure and tensile properties had significant variations, with the appearance of globular pearlite (Pg in Table 3) because of exposure at high temperatures and a moderate increase in grain size. Consequently, there was a decrease in HB, YS and UTS, and consistently the Θ_0 also decreased. After heat treatment at a temperature of 970 °C, the microstructures seemed to restore partially to the original microstructure (summarized in Table 2), which has been reported in [26,27], besides the pearlite grain size that increased significantly after every heat treatment, particularly at a temperature of 970 °C. Consistently, the HB, YS and UTS after austenitizing at a temperature of 970 °C increased significantly with respect to the heat treatments at temperatures of 700 °C and 750 °C, but did not achieve the original values of the as-supplied condition, see Table 2. The apparent fracture toughness decreased for ER7 after heat treatments at temperatures of 700 °C and 750 °C, remained unchanged for HYPERLOS® and Class C, and improved slightly for Class B and SANDLOS®. After heat treatment at a temperature of 970 °C for the steels ER7 and HYPERLOS® K_Q decreased significantly, while it was unchanged for Class B, Class C and SANDLOS® which presented the worst K_Q in the as-supplied condition, showing a certain stability of the apparent fracture toughness. The Voce parameter Θ_o which was inversely related to the fineness of the microstructure (Equation (3)) reduced dramatically after heat treatments at a temperature of 700 °C and 750 °C for all the steels, supporting the microstructure observations in Table 3, and after heat treatment at a temperature of 970 °C seemed to restore the original values of the steels ER7, HYPERLOS® and Class C in the as-supplied condition, while for Class B and SANDLOS® Θ_0 was significantly lower than the as-supplied steels, see Table 2.

In Figure 5, the best fitting Voce parameters of the steels in the as-supplied condition (black open circle data) and after heat treatments (red data) were reported as $1/\varepsilon_c$ vs. Θ_0 in the MAD (Figure 5). In Figure 5, the best fitting red dotted line comes from fitting only the data after heat treatments (red data), and the as-supplied condition data (black open circle data) lay quite close to the best fitting line. Thus, all the data lay close to a single line, confirming the hypothesis suggested in Figure 1 for the steels in the as-supplied condition: the different chemical compositions, production routes and heat treatments had significant effects on the different microstructural constituents of the steels, like the volume fraction of ferrite, pearlite (lamellar or globular), bainite, martensite and pearlite grain size, but there were no significant effects on the dislocation activity. So, the soft ferrite where dislocation motion, multiplication and dynamic recovery occurred was similar in all the steels in the as-supplied condition and after heat treatments, even if the chemical compositions of the steels were different. The applied heat treatment routine increased the presence of soft ferrite and globular pearlite and coarsened the grain size after annealing at temperatures of 700 °C and 750 °C, see Table 3, lowering the values of Θ_0, as reported in Table 3 and Figure 5 (open red square data and open red triangle data, respectively). However, after the heat treatment at a temperature of 970 °C (solid red circle data), the Voce parameter Θ_0 was restored similarly to the as-supplied condition for the steels HYPERLOS® and Class C, even if apparently the microstructure reported in Table 3 did not match the as-supplied microstructure reported in Table 2. In the other steels, namely, ER7, Class B and SANDLOS®, Θ_0 turned out to be smaller than the values from the as-supplied steels, even if apparently the microstructures were restored.

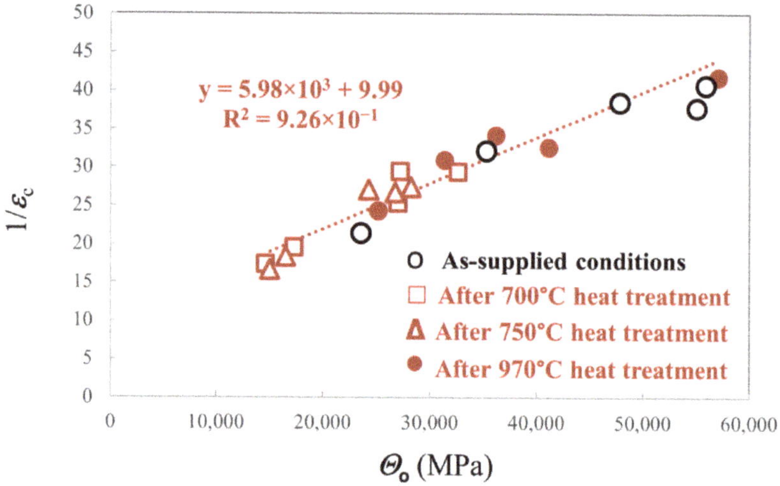

Figure 5. Matrix Assessment Diagram (MAD) of the five steels tensile-tested in the as-supplied condition (open black circle data) and after heat treatments (red data).

In Figure 6, the apparent fracture toughness K_Q values were reported vs. the Voce parameter Θ_0 of the steels (Figure 6) after the heat treatments at temperatures of 700 °C, 750 °C and 970 °C; the data of the steels in the as-supplied condition are reported for comparison's sake. In Figure 6a,b, after annealing at temperatures of 700 °C and 750 °C there were significant reductions in the values of Θ_0, consistently with the coarsening of the microstructures, as supported by microstructure observations reported in Table 3. After annealing at temperatures of 700 °C and 750 °C, the reduction of the apparent fracture toughness was significant for ER7, while K_Q was unchanged for steels HYPERLOS® and Class C with respect to the as-supplied condition, and it even improved slightly for steels Class B and SANDLOS®. After heat treatment at a temperature of 970 °C the microstructures had microstructures similar to the original conditions of the as-supplied steels HYPERLOS® and Class C having some traces of martensite, see Tables 2 and 3, so for these steels the Voce parameter Θ_0 went back to similar values of the as-supplied condition, as reported in Table 3 and Figure 6c, while, in contrast, the steels ER7, Class B and SANDLOS® did not go back to the original conditions. However, after austenitizing at a temperature of 970 °C the apparent fracture toughness of HYPERLOS® and ER7 degraded further in respect to the original values, while the K_Q values of Class B, Class C and SANDLOS® kept similar values to the as-supplied condition. In conclusion, after the heat treatment at a temperature of 970 °C only the steel Class C restored both microstructure and apparent fracture toughness that were like the values of the steel in the as-supplied condition.

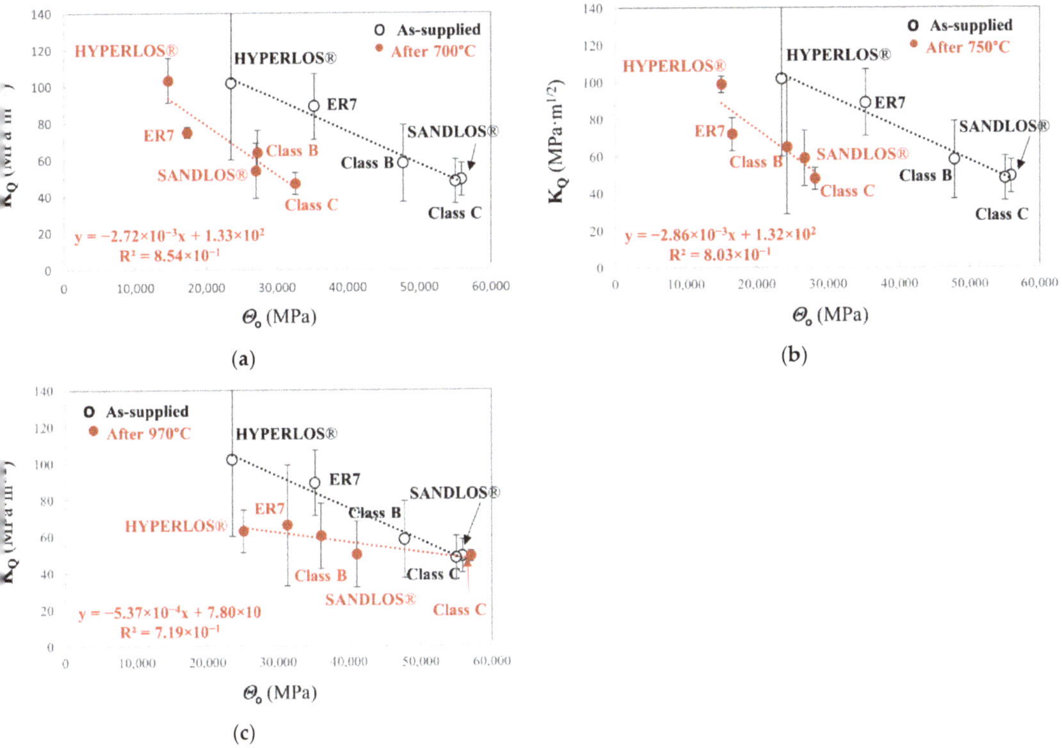

Figure 6. Apparent fracture toughness K_Q vs. best fitting Voce parameter Θ_o of the steels in the as-supplied condition (open black circle data) and after heat treatments (solid red circle data): (**a**) at a temperature of 700 °C; (**b**) at a temperature of 750 °C; (**c**) at a temperature of 970 °C. In all plots, the as-supplied data (open black circle data) are reported for comparison's sake.

4. Discussion

4.1. Steels in the As-Supplied Condition

Microstructure–properties–production relationships are paramount to assess the production route, the potential applications of metallic materials and the material evolution during service. Five railway wheel steels were investigated in the as-supplied condition and after various heat treatments to simulate the microstructural alterations during braking. Besides the conventional mechanical properties, like HB Brinell hardness, YS UTS and A5 presented in [26,27], in the present paper the strain hardening behavior of the steels was investigated, through fitting the true stress and true strain data with the Voce equation, that is, a dislocation-density-related constitutive equation with physical bases. The fitting procedure was based on dealing with the experimental differential tensile data, namely, $d\sigma/d\varepsilon_p$ vs. σ, and by fitting these data with the linear differential Voce equation (Equation (2)) to find the best fitting Voce parameters. It has been reported in [24,25] that, plotting the best fitting Voce parameters $1/\varepsilon_c$ vs. Θ_o in a diagram called Matrix Assessment Diagram (MAD) for different metallic alloys with different chemical compositions and production routes, the Voce parameters lay alongside unique lines that identify unambiguously the materials. The procedure was tuned with foundry products, like spheroidal cast irons, austempered ductile irons (ADIs) and isothermed ductile irons (IDIs). Furthermore, the procedure has been capable of identifying defectiveness in foundry products, where metallurgical discontinuities and defects are particularly copious [28]. The success of using the Voce equation relies on solid physical bases, and the Voce parameters are related to the dislocation dynamics [23]. In Figure 2, the mechanical properties HB, YS and UTS of the steels in the as-supplied condition are plotted vs. the best fitting Voce parameter Θ_0. Linear trends were found, confirming that the Voce parameter Θ_0 was genuinely related to the coarseness of the microstructure, as stated in Equation (3). In fact, the finer the microstructure, the higher the hardness HB, the YS and UTS. So, in Figure 1 in the MAD the best fitting Voce parameters of the five steels in the as-supplied condition are reported, namely, $1/\varepsilon_c$ vs. Θ_0.

The data lay alongside a line, which suggested that the different chemical compositions and production routes had significant effects on the different microstructural constituents that strengthen or soften the steels, but there was no significant effect on the dislocation activity, i.e., on how in the soft ferrite dislocation motion, multiplication and dynamic recovery occurred, even if the chemical compositions of the ferrite were different. The investigated steels are low-alloyed hypoeutectoid steels, where the main alloying elements were C, Mn and Si. Their microstructures came into proeutectoid ferrite and pearlite colonies. Ferrite consisted of BCC crystallographic structure, while pearlite was a eutectoid product coming into alternating lamellar cementite (Fe_3C, orthorhombic crystallographic structure) and lamellar ferrite. Cementite was hard and brittle, while ferrite was soft and tough, and dislocation motion and activity occurred in ferrite. Bainite is also a eutectoid product of austenite transition resulting from faster cooling rates, it consists of fine plate-like cementite in ferrite matrix and it is harder than pearlite. Martensite is extremely hard and brittle, is body-centered tetragonal and results from extremely fast cooling rates. The microstructural constituents like pearlite, bainite and martensite depend mainly on production route that determines the pre-transforming state of the austenite, like austenite grain size, C contents in austenite and cooling rates. So, as the investigated steels had similar production routes [26,27] and heat treatments, C was expected to be relevant in determining the different microstructure, so the Voce parameters Θ_0 that are related to the microstructure, as stated in Equation (3), are plotted vs. the C content in Figure 7. It is evident that with lower C content (ER7 and HYPERLOS®) the microstructure was softer (lower Θ_0), because it was richer in ferrite, while with increasing C content the microstructure was harder (higher Θ_0) with increasing pearlite content. Similar trends could be found with the other mechanical properties, that is, HB, YS and UTS.

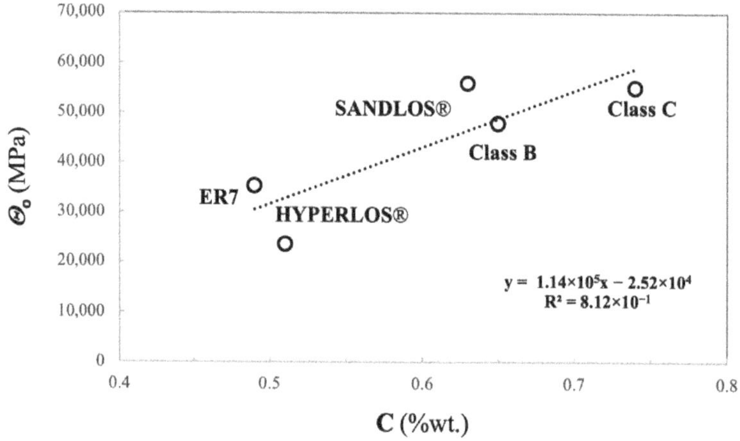

Figure 7. Θ_0 Voce parameter vs. C content (%wt.) for the five steels in as-supplied condition.

ER7 presented the lowest C content (0.49%wt., see Table 1), resulting in the highest ferrite phase fraction: therefore, Θ_0 was low (23,549 MPa) and HB, YS and UTS were low. HYPERLOS® had C, Mn and Si contents quite similar to ER7, while the main difference was the higher V content that formed stable carbides, reducing the pearlite content, and increasing significantly ductility (28%, see Table 2) and apparent toughness (102.2 ± 14 MPa·m$^{1/2}$) with respect to ER7 (14% and 89 ± 6 MPa·m$^{1/2}$). Θ_0 seemed to represent properly this difference by far more than HB, YS and UTS, assuming values of 23,549 MPa and 35,273 MPa for HYPERLOS® and ER7, respectively. Steels Class B and SANDLOS® had intermediate C content (0.65 and 0.63%wt., respectively), producing a volume fraction of ferrite that was intermediate between ER7, HYPERLOS® and Class C that had the highest C content (0.74 wt.%), which produced a pearlitic microstructure with little ferrite. So, in Class C steel HB, YS and UTS were higher, and the Voce parameter Θ_0 assumed a higher value of 55,046 MPa, while, consistently, the apparent fracture toughness decreased to 48 ± 4 MPa·m$^{1/2}$. SANDLOS® is an upgraded Class B modified [26,27] with a higher content of Mn and Si. Mn is a pearlite refiner, reducing the pearlite channels' dimension, and, as a consequence, increasing the HB hardness, YS, UTS and the work-hardening behavior [26,27,34]. Si partitions into the ferrite, strengthening it by a solid solution mechanism, and improving HB and cyclic YS [24,26,34]. So higher Mn and Si contents caused the increase in mechanical properties, like HB, YS and UTS, and the increase in the Voce parameter Θ_0 (55,910 MPa) with respect to Class B, achieving values like steel Class C. Consistently, the apparent fracture toughness was 49 ± 3 MPa·m$^{1/2}$, similar to Class C. In Figure 3, the apparent fracture toughness K_Q values vs. the Voce parameter Θ_0 are reported for the steels in the as-supplied condition. A good linearity was found between K_Q and Θ_0 for the steels, suggesting that the Voce parameter might be a good parameter to index the fracture properties of these steels in the as-supplied condition. Indeed, K_Q was reported also vs. HB hardness, YS and UTS in Figure 4a–c, respectively, where the data shown poorer linear relationships. Furthermore, though also in Figure 4 good linear trends were found, it was noteworthy that the data points for SANDLOS® and Class C did not match, even if fracture toughness was the same, see Tables 2 and 3. In fact, in Figure 3, where K_Q vs. Θ_0 was plotted, the data points of SANDLOS® and Class C overlap perfectly. Indeed, the Voce parameter Θ_0 might be a more reliable parameter than HB, YS and UTS to represent the microstructure and to be related to the apparent fracture toughness. In cases of HB, this could be rationalized by the fact that hardness measurements concern mainly compressive loading, as eventual microstructure features that might nucleate and propagate cracks and fracture were not active during compression. In cases of YS and UTS measurements that presented the best linearities in Figure 4b and Figure 4c, respectively, YS concerns dislocation motion through obstacles (inversely proportional to mean obstacle distance) but also concerned dislocation unlocking because of anchor phenomena of C atoms, which is known as static strain ageing and is described by the Bilby–Cottrell atmosphere theory. UTS comes from the balance between athermal dislocation multiplication hardening and dynamic recovery softening. So, from the strain hardening theory of Kocks and Mecking, the parameter Θ_0 that is related only to the athermal dislocation multiplication phenomenon [23] seemed to be the mechanical parameter related to the microstructure features more relevant for the apparent fracture toughness.

4.2. Steels after Heat Treatments

After heat treatments at temperatures of 700 °C and 750 °C, the microstructure and mechanical properties had significant variations, with the partial dissolution of lamellar pearlite to form into globular pearlite and a moderate increase in grain size, see Table 3 and references [26,27] for more details. Consequently, a decrease in HB, YS, UTS and Θ_0 occurred, as reported in Table 3. After austenitizing at a temperature of 970 °C the microstructures seemed to restore partially to the original microstructures, which was found in [26,27] and is summarized and reported in Table 3, besides grain size which increased considerably. Consistently, the HB, YS and UTS after 970 °C increased significantly with respect to the heat treatments after 700 and 750 °C, without achieving, however, the original values of the as-supplied condition, see Tables 2 and 3 for comparison, and reference [26,27] for detailed microstructure observations. Indeed, only HYPERLOS® and Class C steels had mechanical properties comparable to the as-supplied condition, while for the other steels the properties were slightly lower. The apparent fracture toughness K_Q decreased significantly for ER7 after heat treatments at temperatures of 700 °C and 750 °C, was unchanged for HYPERLOS® and Class C, and increased slightly for steels Class B and SANDLOS®. After the heat treatment at a temperature of 970 °C, the microstructure was similar to the original condition of the as-supplied HYPERLOS® and Class C steels, see Tables 2 and 3, so for these steels the Voce parameter Θ_o went back to similar values of the as-supplied condition, as reported in Table 3. In contrast, the ER7, Class B and SANDLOS® steels did not go back to the original conditions. However, after 970 °C heat treatment the apparent fracture toughness of HYPERLOS® and ER7 degraded further with respect to the original values, while the K_Q values of Class B, Class C and SANDLOS® kept similar values to the as-supplied condition. In conclusion, after the heat treatment at 970 °C only the steel Class C had both similar microstructure and apparent fracture toughness to the values of the steel in the as-supplied condition. Figure 5 shows the best fitting Voce parameters for steels in both the as-supplied condition (black open circles) and after heat treatments (red data) in a $1/\varepsilon_c$ vs. Θ_o plot. The red dotted line, fitted from the heat-treated data, also aligns closely with the as-supplied data, indicating that despite differences in chemical composition, production methods and heat treatments, all data align on a single line. This suggests significant effects on the microstructure components such as ferrite, pearlite, bainite, martensite and pearlite grain size. Conversely, the soft ferrite where dislocation multiplication and dynamic recovery occurred was not substantially affected after heat treatments, even if chemical compositions were different. After heat treatments at temperatures of 700 °C and 750 °C, the Voce parameters $1/\varepsilon_c$ and Θ_o lowered because of the coarsening of the microstructure, while after austenitizing at a temperature of 970 °C they increased assuming values similar to the as-supplied ones, because of refining of the matrix microstructure, while the pearlite grain size was finally coarser. In Figure 6, the apparent fracture toughness K_Q values are plotted vs. the Voce parameter Θ_o of the steels after the heat treatments at temperatures of 700 °C, 750 °C and 970 °C, and the data of the steels for all heat treatment routines and material in as-delivered condition (Figure 6). After high temperature treatments, the data seemed to have good linearity, and in Figure 6c it is evident that the steel Class C was the only one to have a microstructure similar to the original as-supplied condition (microstructure and K_Q), while for the other steels either microstructure (Θ_o) or apparent fracture toughness K_Q were lower after austenitizing at a temperature of 970 °C without any full recovery to the as-supplied condition. Microstructure and K_Q evolution can be rationalized on the critical temperatures of the Fe-Fe_3C state diagram and different chemical compositions of the steels. The temperature of 700 °C was slightly below the eutectoid temperature A_1, so the heat treatment changed the original lamellar pearlite morphology, resulting in ferrite with lamellar and new globular pearlite, and a bigger grain size compared to the as-supplied steels. At a temperature of 750 °C, the steels were heated between the critical temperatures A_1 and A_3, causing a partial austenitization and resulting in a final microstructure also consisting of ferrite with lamellar and new globular pearlite. At a temperature of 970 °C, the steels were heated above A_3, fully austenitizing and resulting after cooling in a microstructure of lamellar pearlite and ferrite for all steels, with bainite and martensite traces for SANDLOS® and Class C steels. Though Si between 400 and 600 °C, and particularly Mn at temperature higher than 600 °C, has stabilizing properties for pearlite, the values of Θ_o in Figure 6a,b were significantly reduced with respect to the as-supplied condition. The reduction was more significant for the harder steels, like SANDLOS® and Class C. Furthermore, after heat treatments at temperatures of 700 °C and 750 °C the apparent fracture toughness K_Q reduced significantly for ER7, but kept quite constant for the other steels, besides Class B and SANDLOS® where K_Q even improved. This finding could be attributed to the appearance of globular pearlite that can positively affect toughness [26,27,34]. Conversely, ER7 that had the lowest content of C, had the highest volume fraction of ferrite, resulting in less influence by the globular pearlite, while HYPERLOS® with similar content of C had vanadium carbides that are stable at low temperatures. After full austenitizing at a temperature of 970 °C the full annealing and cooling restored microstructural constituents with high strength and strain hardening capability, like lamellar pearlite, bainite and martensite, while globular pearlite was not present. The heat treatment at temperatures of 970 °C heat treatment could restore the low K_Q of steels Class B and SANDLOS® of the as-supplied condition, but the apparent fracture toughness of the softer steels ER7 and HYPERLOS® were degraded. The poor K_Q properties of Class C were indeed maintained constant after any heat treatment. So, as reported in Figure 6c, only Class C restored fully the original microstructure and toughness conditions of the as-supplied condition.

4.3. Considerations on the Use of Θ_0 and the Other Mechanical Properties to Relate to K_Q

In Figure 2, the mechanical properties HB, YS and UTS were plotted vs. Θ_0, showing good linearities that supported Equation (3), where Θ_0 was stated to be proportional to the microstructure coarseness. So, the apparent fracture toughness K_Q could be plotted vs. HB, YS and UTS with good linearities, as indeed was found in Figure 4 for the steels in the as-supplied condition. As mentioned in Section 3.1, HB hardness was related to compressive loading, and this explained the worst relationship with K_Q. However, YS and UTS that involve tensile loading had better linear relationships with K_Q. So, in Figure 8 the apparent fracture toughness K_Q vs. YS and UTS of the steels after the heat treatments at temperatures of 700 °C, 750 °C and 970 °C are reported (red solid circle data) to investigate the capability of different tensile parameters to be related to the apparent fracture toughness; in all plots, as-supplied data (open black circle data) were presented for comparison's sake. The information that can be gathered from YS and UTS in Figure 8 was quite equivalent to what was reported in Figure 3 for K_Q vs. Θ_0. The square correlation coefficients appear to be slightly better for K_Q vs. Θ_0, and between YS and UTS the values of R^2 seem to be slightly better for the latter. In Figure 9, a comparison is shown between YS and UTS and the best fitting Voce parameter Θ_0 of the steels in the as-supplied condition (black open circle data) and after heat treatments (red data).

After high temperature heat treatments, good linear relationships between YS, UTS and Θ_0 were found, proving that the Voce parameter Θ_0 was really related to the coarseness of the microstructure (Figure 9). However, it was noteworthy that in the plot YS vs. Θ_0 there was a step between some discontinuities in the trend. The red dotted best fitting lines were calculated on the steel data after heat treatments (red data): in Figure 9a for YS vs. Θ_0, the as-supplied steels data were quite far away from the best fitting line, while in Figure 9b, UTS vs. Θ_0, the as-supplied steels data lay about the red dotted best fitting line. This finding could be rationalized by the fact that, even if YS and UTS concern tensile loading, yielding dealt with dislocation motion and also unlocking of dislocations from a C atoms atmosphere that anchors dislocations, a phenomenon known as strain ageing and described by the Bilby–Cottrell atmosphere theory [35]. When unlocking occurred at yield showing a sharp yield point, plastic deformation began and the C atoms did not interact any more with dislocation motion and activity, unless tensile tests were at intermediate temperatures where the interaction between dislocations and solute atoms might be dynamic, giving rise to dynamic strain ageing, also known as the Portevin–Le Chatelier effect [34]. So, the sharp yield point was a static phenomenon, that was dependent on the initial solute atoms' distribution, but does not affect strain hardening or eventually fracture behavior. Strain ageing with sharp yield point occurred for steels at room temperature depending on the chemical compositions and production routes, and the investigated steels in the as-supplied condition and after heat treatments did not always present sharp yield points, as is summarized in Table 4. This finding should explain the data in Figure 9, and at the same time, it suggests that UTS and Θ_0 should be the best parameters to describe the effective microstructure of the investigated steels. So, the use of diagram K_Q vs. Θ_0 or

K_Q vs. UTS should be encouraged, even if the results coming from using K_Q vs. YS also appeared meaningful, since they present good linearities.

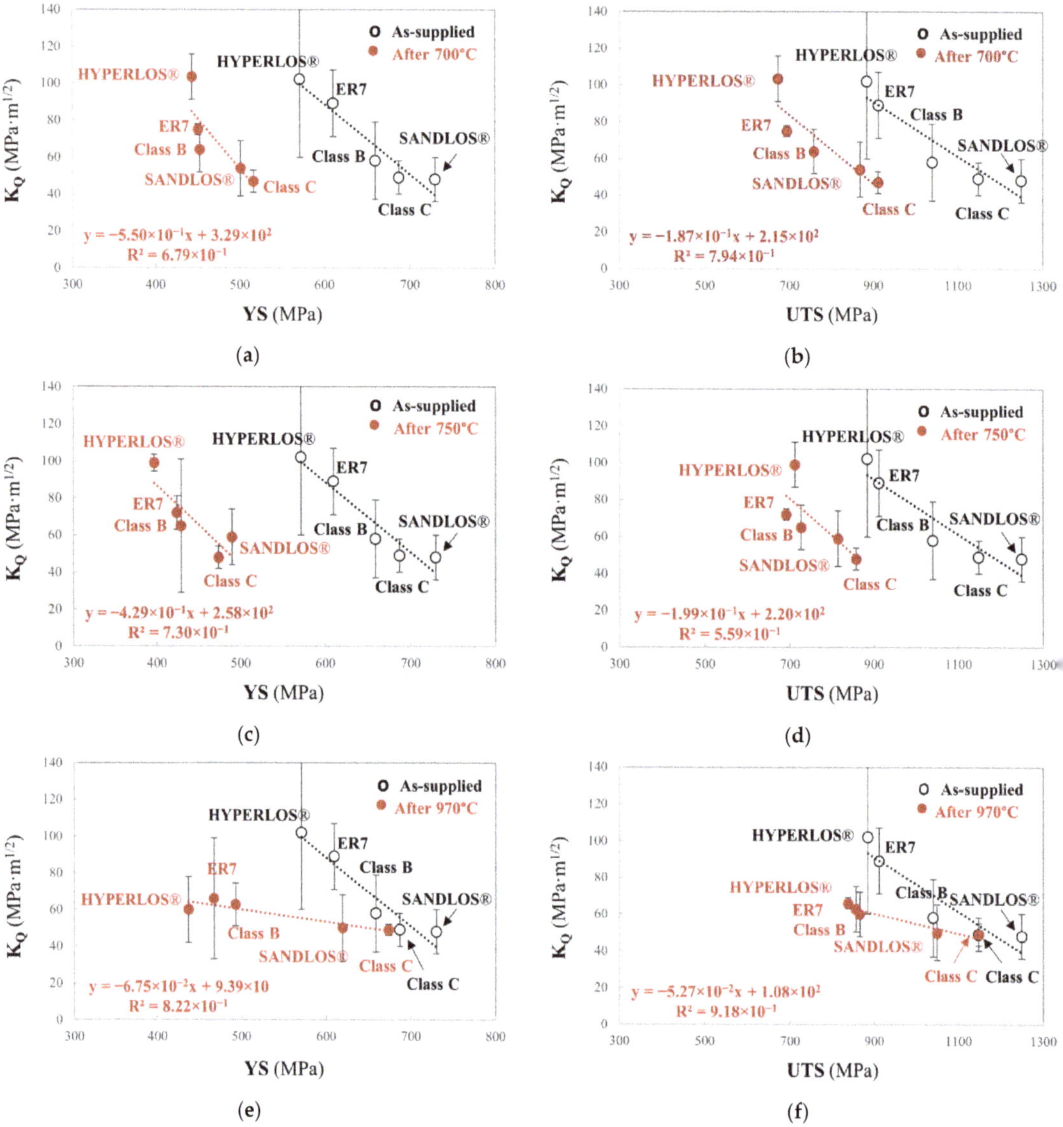

Figure 8. Apparent fracture toughness K_Q vs. yield stress (YS) of the steels in the as-supplied condition (open black circle data) and after heat treatments (solid red circle data): (**a**) at a temperature of 700 °C; (**c**) at a temperature of 750 °C; (**e**) at a temperature of 970 °C. Apparent fracture toughness K_Q vs. ultimate tensile stress (UTS) of the steels in the as-supplied condition (open black circle data) and after heat treatments (solid red circle data): (**b**) at a temperature of 700 °C; (**d**) at a temperature of 750 °C; (**f**) at a temperature of 970 °C. In all plots, the as-supplied data (open black circle data) are reported for comparison's sake.

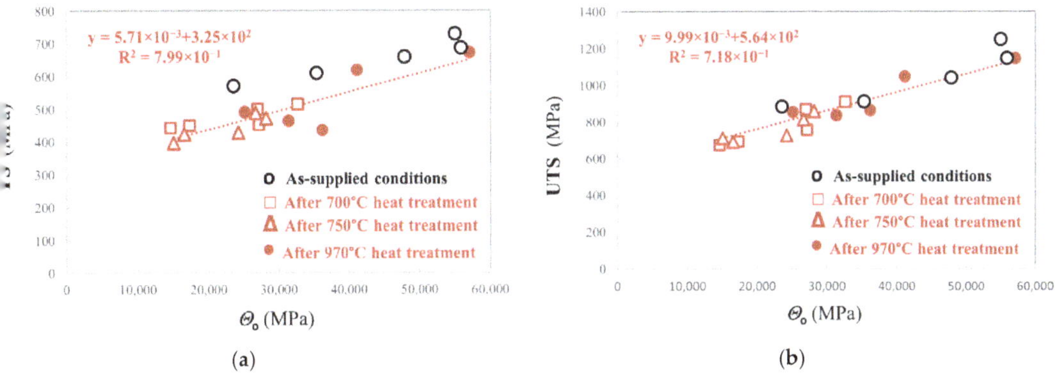

Figure 9. Comparison between selected mechanical properties and the best fitting Voce parameter Θ_o of the steels in the as-supplied condition and after heat treatments: (**a**) yield stress YS vs. Θ_o; (**b**) ultimate tensile stress UTS vs. Θ_o.

Table 4. Sharp yield point appearance in the tensile curves of the investigated steels in the as-supplied condition and after heat treatments.

Steel	Sharp Yield Point			
	As-Supplied	After 700 °C	After 750 °C	After 970 °C
ER7	YES	YES	YES	NO
HYPERLOS®	YES	YES	YES	NO
Class B	YES	YES	YES	YES
SANDLOS®	NO	YES	YES	NO
Class C	NO	NO	YES	NO

5. Conclusions

The quality-assessment procedure based on strain hardening and the Voce equation was used to analyze the results of experimental hardness, tensile and toughness tests. These tests involved five railway wheel steels in the as-supplied condition and after three different heat treatments designed to replicate the modification of the microstructure due to braking. The correlation between microstructure features of the steels and the best fitting Voce parameters was reported, as well as the relation with Brinell hardness measurements (HB) and apparent fracture toughness (K_Q) in controlled conditions. The quality-assessment procedure based on strain hardening and the Voce equation supported the evidence that investigated steels had a different microstructure and mechanical stability when undergoing severe thermal cycles. The following conclusions could be drawn:

- The Voce parameter Θ_o was genuinely related to the coarseness of the microstructure, and linear relationships with HB, YS and UTS were found;
- In the Matrix Assessment Diagram MAD ($1/\varepsilon_c$ vs. Θ_o), the as-supplied steel data lay alongside a line, which suggested that the different chemical compositions and production routes had significant effects on the different microstructural constituents, but ferrite where dislocation activity occurred was the same for the five steels;
- The apparent fracture toughness K_Q vs. the Voce parameter Θ_o presented good linearities for the steels in the as-supplied condition, suggesting that the Voce parameter might be a good parameter to indicate the fracture properties of these steels in the as-supplied conditions;
- The apparent fracture toughness K_Q vs. the Voce parameter Θ_o presented good linearities for the steels after heat treatments, indicating the fracture properties of these steels also after working conditions;

- Also, K_Q vs. YS and K_Q vs. UTS could give valuable information, equivalent to the use of the Voce analysis approach (K_Q vs. YS); however, the physical bases of the Voce parameter can give some additional information.

After heat treatments at temperatures of 700 °C and 750 °C, the properties of the steels decreased, which was consistent with the evolution of the microstructure, while after exposure at a temperature of 970 °C with subsequent cooling in air, Class C steel appeared to have a microstructure similar to the original microstructure, with tensile and toughness properties very similar to the as-supplied conditions, demonstrating better microstructural stability compared to steels ER7, HYPERLOS®, Class B and SANDLOS®.

Author Contributions: Conceptualization, G.A., M.F. and L.G.; methodology, G.A.; validation, M.F. and L.G.; formal analysis, G.A.; investigation, M.F. and L.G.; resources, M.F.; writing—original draft preparation, G.A., M.F. and L.G; writing—review and editing, G.A., M.F. and L.G. All authors have read and agreed to the published version of the manuscript.

Funding: This research received no external funding.

Data Availability Statement: The original contributions presented in the study are included in the article, further inquiries can be directed to the corresponding author.

Acknowledgments: The authors wish to express their gratitude to Michele Nodari, Ivan Meloni, and the staff of the Lucchini RS Metallurgy Department for their support.

Conflicts of Interest: The authors declare no conflicts of interest.

References

1. Landström, E.V.; Vernersson, T.; Lundén, R. Analysis and Testing of Tread Braked Railway Wheel—Effects of Hot Spots on Wh Performance. *Int. J. Fatigue* **2024**, *180*, 108116. [CrossRef]
2. Handa, K.; Ikeuchi, K.; Morimoto, F. Temperature-Dependent Wear of Tread-Braked Railway Wheels. *Wear* **2020**, *452–453*, 2032 [CrossRef]
3. Wu, Q.; Spiryagin, M.; Cole, C. Block–Wheel–Rail Temperature Assessments Via Longitudinal Train Dynamics Simulations *Comput. Nonlinear Dyn.* **2022**, *17*, 111007. [CrossRef]
4. Landström, E.V.; Steyn, E.; Ahlström, J.; Vernersson, T. Thermomechanical Testing and Modelling of Railway Wheel Steel. *Int Fatigue* **2023**, *168*, 107373. [CrossRef]
5. Haidari, A.; Hosseini-Tehrani, P. Fatigue Analysis of Railway Wheels Under Combined Thermal and Mechanical Loads. *J. The Stress.* **2014**, *37*, 34–50. [CrossRef]
6. Nikas, D.; Ahlström, J.; Malakizadi, A. Mechanical Properties and Fatigue Behaviour of Railway Wheel Steels as Influenced Mechanical and Thermal Loadings. *Wear* **2016**, *366–367*, 407–415. [CrossRef]
7. Cvetkovski, K.; Ahlström, J.; Karlsson, B. Thermal Degradation of Pearlitic Steels: Influence on Mechanical Properties Includi Fatigue Behaviour. *Mater. Sci. Technol.* **2011**, *27*, 648–654. [CrossRef]
8. Ahlström, J.; Karlsson, B. Microstructural Evaluation and Interpretation of the Mechanically and Thermally Affected Zone unc Railway Wheel Flats. *Wear* **1999**, *232*, 1–14. [CrossRef]
9. Allain, S.Y.P.; Roth, A.; Bouaziz, O.; D'Eramo, E. Microstructure-Based Behavior Law for Globular Pearlitic Steels. *J. Mater. R Technol.* **2019**, *8*, 3373–3376. [CrossRef]
10. Ghidini, L.; Faccoli, M.; Zani, N.; Petrogalli, C.; Bonometti, S.; Mazzu, A. An Innovative Small-Scale Testing Procedure to Stu Damage in Shoe-Braked Wheels. *Proc. Inst. Mech. Eng. F J. Rail Rapid Transit* **2023**, *238*, 414–426. [CrossRef]
11. Jergéus, J. Martensite Formation and Residual Stresses around Railway Wheel Flats. *Proc. Inst. Mech. Eng. C J. Mech. Eng. S* **1998**, *212*, 69–79. [CrossRef]
12. Handa, K.; Kimura, Y.; Mishima, Y. Ferrite and Spheroidized Cementite Ultrafine Microstructure Formation in an Fe-0.67 Pct Steel for Railway Wheels under Simulated Service Conditions. *Metall. Mater. Trans. A* **2009**, *40*, 2901–2908. [CrossRef]
13. Petersson, M.; Vernersson, T. Noise-Related Roughness on Tread Braked Railway Wheels-Experimental Measurements a Numerical Simulations. *Wear* **2002**, *253*, 301–307. [CrossRef]
14. Vernersson, T. Thermally Induced Roughness of Tread-Braked Railway Wheels: Part 1: Brake Rig Experiments. *Wear* **1999**, 2 96–105. [CrossRef]
15. Vernersson, T. Thermally Induced Roughness of Tread Braked Railway Wheels: Part 2: Modelling and Field Measurements. *W* **1999**, *236*, 106–116. [CrossRef]
16. Faccoli, M.; Provezza, L.; Petrogalli, C.; Ghidini, A.; Mazzù, A. A Small-Scale Experimental Study of the Damage Due Intermittent Shoe Braking on the Tread of High-Speed Train Wheels. *Tribol. Trans.* **2020**, *63*, 1041–1050. [CrossRef]

Faccoli, M.; Zani, N.; Ghidini, A.; Petrogalli, C. Experimental and Numerical Investigation on the Wear Behavior of High Performance Railway Wheel Steels Paired with Various Brake Block Materials under Dry Sliding Conditions. *Wear* **2022**, *506–507*, 204456. [CrossRef]
Faccoli, M.; Zani, N.; Ghidini, A.; Petrogalli, C. Tribological Behavior of Two High Performance Railway Wheel Steels Paired with a Brake Block Cast Iron. *Tribol. Trans.* **2021**, *65*, 296–307. [CrossRef]
Vernersson, T. Temperatures at Railway Tread Braking. Part 2: Calibration and Numerical Examples. *Proc. Inst. Mech. Eng. F J. Rail Rapid Transit.* **2007**, *221*, 429–441. [CrossRef]
Handa, K.; Kimura, Y.; Mishima, Y. Surface Cracks Initiation on Carbon Steel Railway Wheels under Concurrent Load of Continuous Rolling Contact and Cyclic Frictional Heat. *Wear* **2010**, *268*, 50–58. [CrossRef]
Orringer, O.; Gray, D.E. Thermal Cracking in Railroad Vehicle Wheels Subjected to High Performance Stop Braking. *Theor. Appl. Fract. Mech.* **1995**, *23*, 55–65. [CrossRef]
Caprioli, S.; Vernersson, T.; Handa, K.; Ikeuchi, K. Thermal Cracking of Railway Wheels: Towards Experimental Validation. *Tribol. Int.* **2016**, *94*, 409–420. [CrossRef]
Kocks, U.F.; Mecking, H. Physics and Phenomenology of Strain Hardening: The FCC Case. *Prog. Mater. Sci.* **2003**, *48*, 171–273. [CrossRef]
Angella, G.; Donnini, R.; Zanardi, F. Assessment of Microstructure Effects on Tensile Behaviour in Silicon Strengthened Ductile Irons Produced through Different Cooling Rates. *Int. J. Cast Met. Res.* **2020**, *33*, 89–102. [CrossRef]
Angella, G.; Cova, M.; Bertuzzi, G.; Zanardi, F. Soundness Discrimination in Ferrite Ductile Irons Through Tensile Data Analysis. *Int. J. Met.* **2020**, *14*, 816–826. [CrossRef]
Faccoli, M.; Ghidini, A.; Mazzù, A. Experimental and Numerical Investigation of the Thermal Effects on Railway Wheels for Shoe-Braked High-Speed Train Applications. *Metall. Mater. Trans. A* **2018**, *49*, 4544–4554. [CrossRef]
Faccoli, M.; Ghidini, A.; Mazzù, A. Changes in the Microstructure and Mechanical Properties of Railway Wheel Steels as a Result of the Thermal Load Caused by Shoe Braking. *Metall. Mater. Trans. A* **2019**, *50*, 1701–1714. [CrossRef]
EN 13262; Railway applications—Wheelsets and Bogies—Wheels—Product Requirements. CEN: Bruxelles, Belgium, 2020.
Angella, G.; Taloni, M.; Górny, M.; Tarasiuk, J.; Wronski, S.; Montanari, R.; Pedranz, M.; Benedetti, M.; Fontanari, V.; Lusuardi, D. An Insight into the Defects-Driven Plasticity in Ductile Cast Irons. *Materials* **2023**, *16*, 3748. [CrossRef]
Angella, G.; Zanardi, F. Defects-Driven Plasticity: Rationalization Based on Notch Strengthening and Continuous Damage Mechanics. *Solid State Phenom.* **2023**, *353*, 169–175. [CrossRef]
ASTM E399; Standard Test Method for Linear-Elastic Plane-Strain Fracture Toughness of Metallic Materials. American Society for Testing and Materials International: West Conshohocken, PA, USA, 2022.
ASTM E112; Standard Test Methods for Determining Average Grain Size. American Society for Testing and Materials International: West Conshohocken, PA, USA, 2021.
Singh, U.P.; Popli, A.M.; Jain, D.K.; Roy, B.; Jha, S. Influence of Microalloying on Mechanical and Metallurgical Properties of Wear Resistant Coach and Wagon Wheel Steel. *J. Mater. Eng. Perform.* **2003**, *12*, 573–580. [CrossRef]
Zhang, Y.; Lu, L.; Gong, Y.; Zhang, J.; Zeng, D. Fretting Wear-Induced Evolution of Surface Damage in Press-Fitted Shaft. *Wear* **2017**, *384–385*, 131–141. [CrossRef]
Abbaschian, R.; Reed-Hill, R.E. *Physical Metallurgy Principles*; Cengage Learning: Stamford, CT, USA, 2008; ISBN 9780495082545.

sclaimer/Publisher's Note: The statements, opinions and data contained in all publications are solely those of the individual hor(s) and contributor(s) and not of MDPI and/or the editor(s). MDPI and/or the editor(s) disclaim responsibility for any injury to ple or property resulting from any ideas, methods, instructions or products referred to in the content.

metals

Article

Effect of Si Content on Microstructure and Properties of Low-Carbon Medium-Manganese Steel after Intercritical Heat Treatment

Zihan Hu and Hanguang Fu *

School of Materials Science and Engineering, Beijing University of Technology, Beijing 100124, China; huzihan@emails.bjut.edu.cn
* Correspondence: hgfu@bjut.edu.cn; Tel.: +86-10-67396093; Fax: +86-10-67396244

Abstract: The microstructure and mechanical properties of three kinds of low-carbon medium-manganese steels with different Si contents under an intercritical heat treatment process were studied. The results show that the microstructure of the test forged steel is mainly composed of ferrite and pearlite. After 900 °C complete austenitizing quenching + 720 °C intercritical quenching, the microstructure of the test steel is mainly composed of ferrite and martensite. With the increase in Si content, the microstructure becomes finer and more uniform. The microstructure of the test steel after 900 °C complete austenitizing quenching + 720 °C intercritical quenching + 680 °C intercritical tempering is dominated by ferrite and tempered martensite, with a small amount of retained austenite and cementite. As the Si content increases, the boundaries between ferrite and tempered martensite become more clear. The tensile strength and hardness of the test steel increase with the increase in Si content, while the elongation first increases and then decreases; the comprehensive performance of the test steel is the best when the Si content is 0.685 wt. %, with a tensile strength of 726 MPa, a yield ratio of only 0.65, the highest elongation of 30.5%, and the highest strong plastic product of 22,143 MPa.%.

Keywords: intercritical heat treatment; Si content; low-carbon medium-manganese steel; microstructure; mechanical properties

1. Introduction

The demand for low-cost, high-strength, and high-toughness low-carbon alloy steel in the automotive, power, machinery, and metallurgical industries is increasing [1,2]. Research has found that a certain amount of metastable austenite in steel can significantly improve its mechanical properties while significantly improving its plasticity and toughness [3,4]. Metastable austenite includes both retained austenite and reverse-transformed austenite. Austenite that remains in the structure without undergoing martensitic transformation during cooling is called retained austenite, and austenite that is regenerated by the inverse transformation of martensite or bainite during tempering is called reverse-transformed austenite [5].

In the 1970s, Miller [6] first proposed a low-carbon medium-manganese steel with a chemical composition of 0.11C-5.7Mn (wt. %) and obtained metastable austenite with a stability of 20–40% in the steel through intercritical heat treatment. Morris et al. [7] found that the steel with a chemical composition of 0.04C-5Mn-0.2Mo (wt. %) had good toughness at −196 °C after quenching and intercritical heat treatments, which was attributed to the refinement of the microstructure and the metastable austenite formed during intercritical annealing. Lee et al. [8] showed that the intercritical heat treatment process can obtain reverse-transformed austenite at room temperature for 3–10% Mn or Ni steel, but the effect is not obvious for low-carbon steel with a Mn content less than 3%.

In order to obtain a high content and metastable austenite in low-carbon low-alloy steel, Xie et al. [9] proposed a new three-step heat treatment process. This process involves heating the steel to near the Ac_1 temperature after the two-step intercritical heat treatment, allowing the austenite to further enrich carbon and alloy elements. Xi et al. [10] conducted a three-step heat treatment on low-carbon copper-bearing steel, obtaining 14.7% retained austenite, with a tensile strength of 900 MPa, an overall elongation increased from 13.2% to 22.2%, and an impact energy at −40 °C increased from 100 J to 219 J. Huang et al. [11] conducted a three-step heat treatment on 0.12 C-3.0 Mn low-carbon medium-manganese steel, obtaining a maximum retained austenite content of 15%, a total elongation exceeding 30%, and an impact energy at −80 °C reaching 200 J. Li et al. [12] conducted a three-step heat treatment on low-alloy TRIP steel, finding that the retained austenite content of the test steel increased significantly, and the yield strength slightly decreased with increasing tempering temperature, but the plasticity and toughness significantly improved with increasing tempering temperature.

Adding alloy elements is a method to adjust the structure and properties of medium-manganese steel. Elements such as C, Mn, Si, Al, Ni, and Ti can affect the structure and properties of medium-manganese steel. C, as a solid solution strengthening element and austenite-stabilizing element, has a significant impact on the microstructure and properties of medium-manganese steel. As the content of C increases, the solid solution strengthening effect and the enhancement in austenite stability increase [13]. Mn is the most important element for stabilizing austenite, and the content of austenite in the structure increases with the increase in Mn content [14]. The addition of the Al element can promote the formation of ferrite and improve the plasticity of medium-manganese steel [15]. Cai et al. [16] added the Ti element to medium-manganese steel and found that the grain size of the steel was refined, the hardness was higher, and the wear resistance was better. Zhang et al. [17] added the Ni element to medium-manganese steel, and through the combination of TRIP and NiAl precipitation mechanisms, the steel had excellent tensile strength (961 MPa) and elongation (53%). The increase in Si content can inhibit the precipitation of carbides, promote the enrichment of the C element in austenite, delay the decomposition process of austenite, and improve the stability of residual austenite [18,19]. Si can also refine the grain size and improve the tensile strength and elongation of steel [20]. Song et al. [21] studied the effect of Si on martensite in medium-manganese steel and found that the increase in dislocation density by Si significantly improved the tensile strength of the steel, and the formation of the dislocation cell structure gave the steel good elongation. At present, no researchers have studied the effect of Si content on the intercritical heat treatment of low-carbon medium-manganese steel.

Si is a common alloying element in steel and is inexpensive. In order to improve the strength and toughness of low-carbon medium-manganese steel, this article studies low-carbon medium-manganese steel with different Si contents through intercritical heat treatment to explore the evolution of the microstructure and mechanical properties of low-carbon medium-manganese steel with different Si contents.

2. Materials and Methods

2.1. Material Casting and Forging

The test steel was melted in a vacuum induction melting furnace under argon protection, using industrial pure iron, electrolytic manganese, 75% Si iron, high-purity graphite, etc. The metal mold was poured into a φ70 mm × 200 mm ingot, and the chemical composition of the test steel was determined using an Optical Emission Spectrometer, as shown in Table 1. The ingot was repeatedly thickened and forged into a plate-like blank with a thickness of 30 mm, with a forging ratio of 5. The initial forging temperature was 1200 °C, the final forging temperature was 920 °C, and it was air-cooled to room temperature. The size of the forged steel plate was 165 mm × 130 mm × 30 mm, and all samples were taken from the forged steel plate.

Table 1. The chemical composition of the tested steel (wt. %).

Sample	C	Si	Mn	P	S	Fe
0.2 Si	0.107	0.180	3.010	0.008	0.009	Bal.
0.7 Si	0.111	0.685	3.174	0.008	0.009	Bal.
1.2 Si	0.110	1.206	3.154	0.008	0.009	Bal.

2.2. *Material Heat Treatment*

The CCT curves of the test steels with different Si contents calculated using JMatPro are shown in Figure 1. Table 2 was made based on the calculation results of JMatPro. It shows the austenite transformation end temperature Ac_3, pearlite-to-austenite transformation temperature Ac_1, and martensite start transformation temperature M_s and end transformation temperature M_f of the test steels with different Si contents calculated using JMatPro. From the table, as the Si content increases, the austenite transformation temperature gradually increases, while the martensite transformation temperature gradually decreases. This is because Si, as a non-carbide-forming element, usually exists in the form of solid solution atoms in ferrite and austenite, which increases the activity of carbon and reduces the activity of iron atoms, thereby increasing the diffusion activation energy and requiring higher temperatures to achieve austenitization [22]. As the Si content increases, the transformation range of ferrite in the test steel increases, while the transformation range of bainite decreases, indicating that when the cooling rate is between 100 and 0.01 °C/s, the start transformation temperature of ferrite gradually increases, and the start transformation temperature of bainite gradually decreases with the increase in Si content. This is because during the austenitization process, the accumulation of Si in ferrite increases the chemical potential of C in ferrite, thereby increasing the driving force for C element diffusion into austenite, effectively improving austenite stability while avoiding the formation of pearlite and bainite during cooling, making it easier to obtain martensite [23].

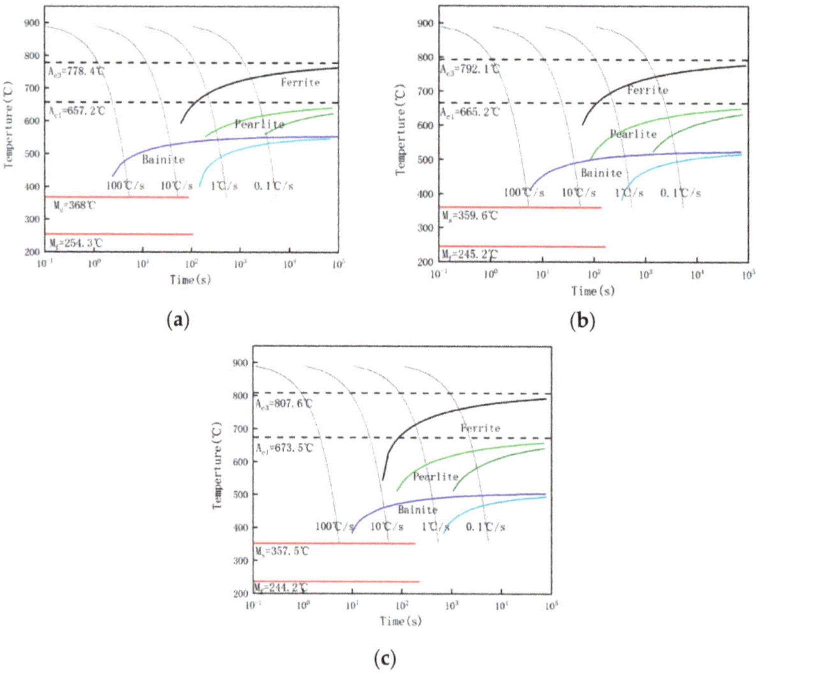

Figure 1. CCT curves of steel tested with different Si contents: (**a**) 0.2 Si; (**b**) 0.7 Si; (**c**) 1.2 Si.

Table 2. Transformation temperature of experimental steel.

Sample	Ac_3 (°C)	Ac_1 (°C)	M_s (°C)	M_f (°C)
0.2 Si	778.4	657.2	368	254.3
0.7 Si	792.1	665.2	359.6	245.2
1.2 Si	807.6	673.5	357.5	244.2

The specific heat treatment process is shown in Figure 2. First, the billet is heated to 900 °C for 30 min for the complete austenitizing heat treatment. Then, the intercritical quenching treatment is carried out. To obtain ferrite's and martensite's dual-phase structure, an appropriate temperature should be set. A lower temperature will lead to excessive ferrite, a higher temperature, and higher martensite content. The temperature of the two-phase zone is 720 °C. The sample is heated to 720 °C for 20 min and water-cooled to room temperature, then heated to 680 °C for 20 min for tempering.

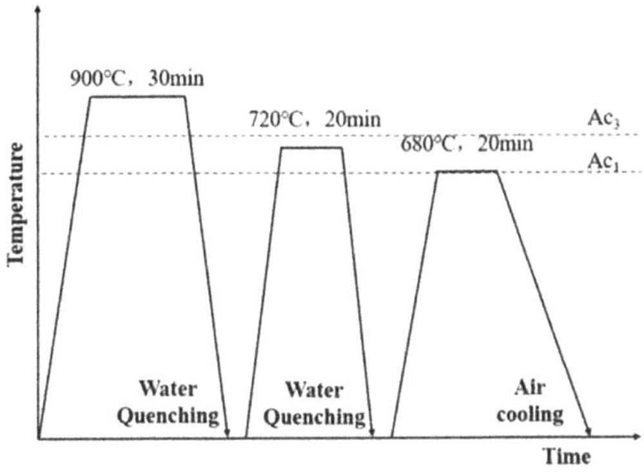

Figure 2. Schematic diagram of heat treatment process of tested steel.

2.3. Microstructure Observation and Mechanical Property Testing

According to the standard GB/T 228.1-2021 "Metallic materials-Tensile testing-Part 1: Method of test at room temperature" [24], the heat-treated specimens were processed into plate-shaped tensile specimens with a total length of 110 mm and a gauge length of 30 mm. The tensile properties were tested at room temperature on a UTM5305 electronic universal testing machine at a tensile rate of 3 mm/min, and the average value of the three specimens was taken as the performance.

The Brinell hardness of the samples was measured according to the standard GB/T 231.1-2018 "Metallic materials-Brinell hardness test-Part 1: Test method" [25]. The diameter of the hard alloy ball was 2.5 mm, the nominal value of the test force was 1839 N, and the ratio of the test force to the ball diameter squared was 30 N/mm^2. Take a sample with a size of 10 mm × 10 mm × 20 mm as a hardness testing sample; four positions were selected for each sample for measurement.

The heat-treated samples were sequentially ground with 80-mesh, 240-mesh, 400-mesh, 1000-mesh, 1500-mesh, and 2000-mesh SiC sandpaper, followed by mechanical polishing with diamond polishing paste with a particle size of W2.5. After polishing, they were etched using a 4% nitric acid alcohol reagent, and microstructure observation and analysis were performed using a field-emission scanning electron microscope (ZEISS Gemini Sigma 500, Carl Zeiss AG, Oberkochen, Germany).

3. Results and Discussions

3.1. Microstructure and Hardness of Forged Plate of Test Steel

Figure 3 shows that the microstructure of the tested forged steel is mainly composed of ferrite and pearlite, which is a typical structure of low-carbon low-alloy steel. In Figure 4, the average values of the Brinell hardness of the tested steel are 279 HB, 315 HB, and 329 HB, indicating an increase in hardness. Si can inhibit the precipitation of carbides, and as the Si content increases, the carbides gradually decrease, and the ferrite becomes finer and more uniform. Si has a high solid solubility in iron, and dissolving in ferrite can significantly strengthen the ferrite and improve the strength and hardness of the steel.

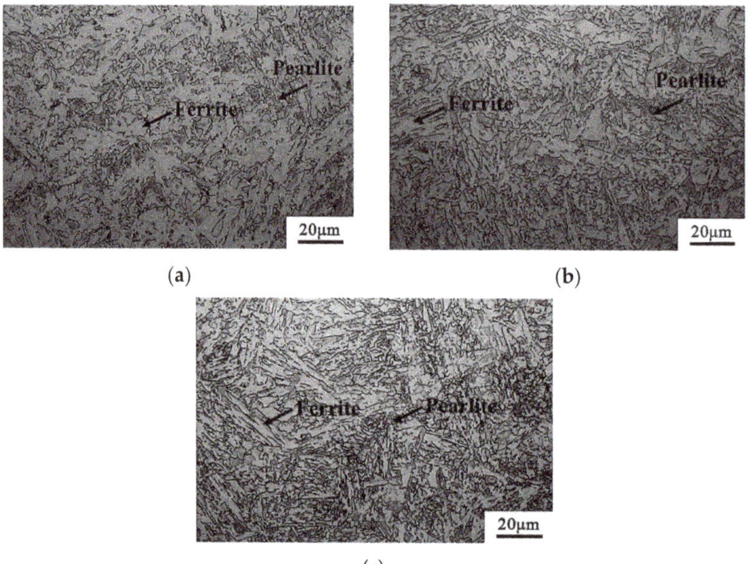

Figure 3. Microstructure of test steel forged plate: (**a**) 0.2 Si; (**b**) 0.7 Si; (**c**) 1.2 Si.

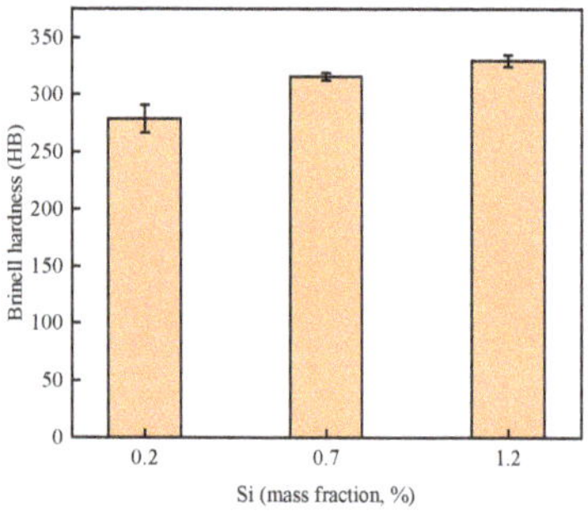

Figure 4. Brinell hardness of test steel forged plate.

3.2. Microstructure and Hardness of Test Steel after Two-Step Heat Treatment

Figure 5 shows the microstructure of the experimental steel after quenching at 900 °C and intercritical quenching at 720 °C. In Figure 5, the microstructure of steel with different Si contents after the two-step heat treatment is mainly composed of ferrite and martensite. With the increase in Si content, the ferrite content in the microstructure decreases, and the ferrite and martensite structures become finer and more evenly distributed, which is conducive to improving the hardness and strength of the steel. In Figure 6, the average values of the Brinell hardness of the tested steel after two steps of heat treatment are 251 HB, 255 HB, and 257 HB, indicating an increase in hardness. In the process of two-phase zone quenching, a reverse transformation austenite structure is formed at the original austenite grain boundary and between lath martensitic and along with the gradual enrichment of austenite-stabilizing elements C and Mn into the reverse transformation austenite, which leads to the improvement in the stability of austenite. In the subsequent cooling process, part of this reverse transformation austenite is transformed into bainite–martensite, and the other part can still stably exist to form a small amount of retained austenite [26].

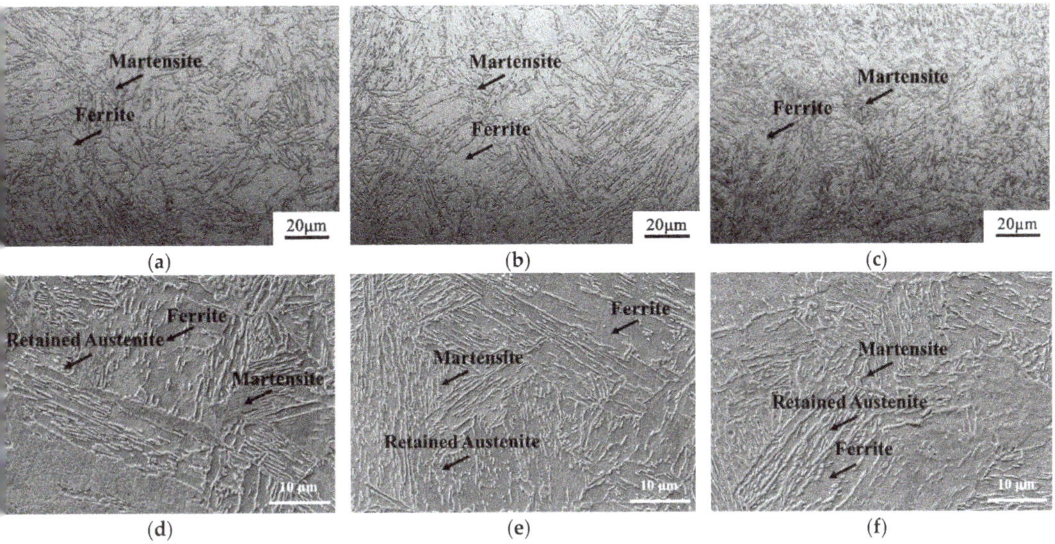

Figure 5. The metallographic microstructure and SEM morphology of the test steel with different Si contents under the annealing process: (**a**,**d**) 0.2 Si; (**b**,**e**) 0.7 Si; (**c**,**f**) 1.2 Si. (**a**–**c**) OM; (**d**–**f**) SEM.

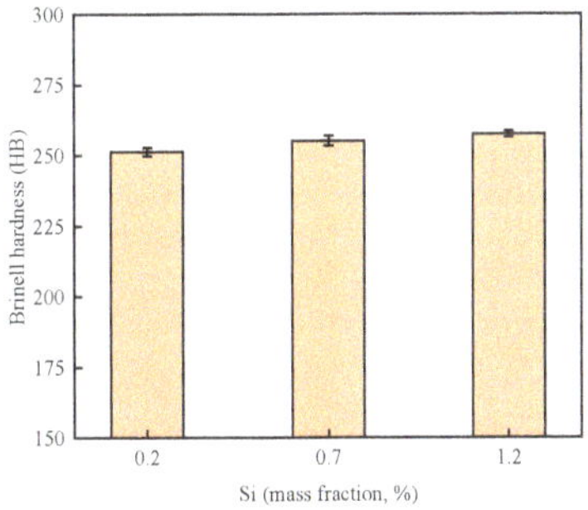

Figure 6. The Brinell hardness of the test steel with different Si contents under the annealing process.

3.3. Microstructure and Mechanical Properties of Test Steel after Intercritical Tempering Treatment

Figure 7 shows the microstructure of the experimental steel after quenching at 900 °C + intercritical quenching at 720 °C + and intercritical tempering at 680 °C. The microstructure of the steel after three-step heat treatment is mainly composed of a dual-phase structure of ferrite and tempered martensite, with a small amount of retained austenite. With the increase in Si content, the boundary of ferrite and tempered martensite becomes clear, and the ferrite matrix has no obvious change. The structure of tempered lath martensite becomes clear with the increase in Si content, and compared with the microstructure after intercritical quenching, the boundary of ferrite and tempered martensite becomes clearer. A large number of dispersed lamellar and fine granular structures can be observed at the ferrite grain boundary or within the grain. These fine particles are mainly cementite formed after high-temperature tempering and finally stable retained austenite [12].

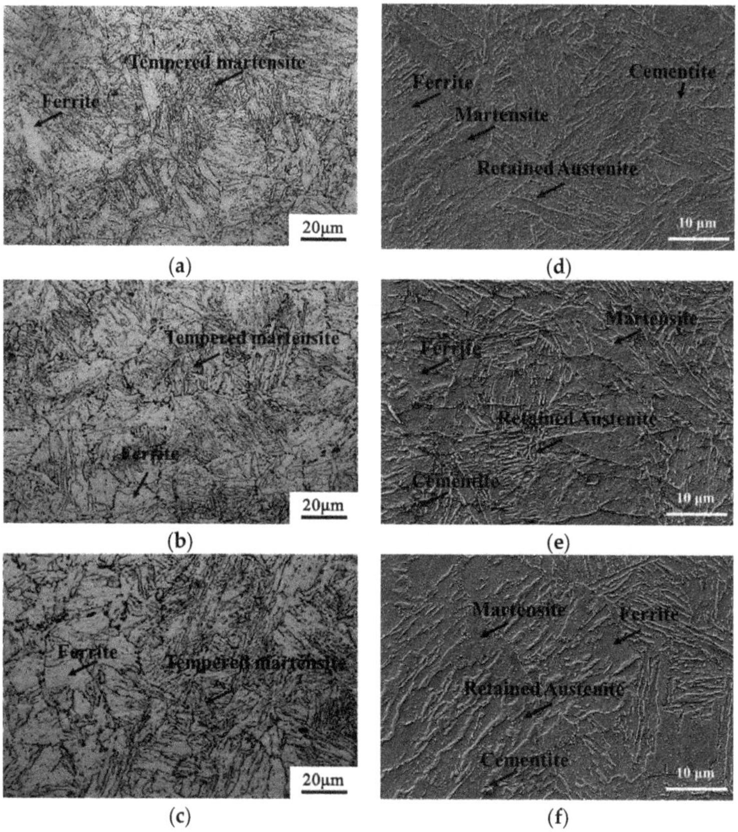

Figure 7. The metallographic microstructure and SEM morphology of the test steel with different Si contents under the tempering process: (**a**,**d**) 0.2 Si; (**b**,**e**) 0.7 Si; (**c**,**f**) 1.2 Si. (**a**–**c**) OM; (**d**–**f**) SEM.

Figure 8 shows the engineering stress–strain curves of samples with different Si contents after three-step heat treatment, and their corresponding mechanical properties are shown in Table 3. Combining Figure 8 and Table 3, the tensile curves of the tested steel after three-step heat treatment are all continuous yield, without an obvious yield plateau, indicating that the plastic deformation of the tested steel mainly occurs during the entire tensile stage, rather than at a specific yield point. As the Si content increases, the tensile strength and hardness gradually increase, the average values of the Brinell hardness are 214 HB, 227 HB, and 234 HB in Figure 9. When the Si content is 1.206 wt. %, the tensile strength is the highest, and the elongation ratio first increases and then decreases. When the Si content is 0.685 wt. %, the elongation is the best.

Table 3. Tensile properties of test steels after intercritical thermal treatment.

Sample	YS (MPa)	TS (MPa)	Y/T Ratio	TEL (%)	PSE (MPa·%)	Hardness (HB)
0.2 Si	480 ± 7	652 ± 3	0.74	28.4 ± 0.3	18,517	214
0.7 Si	474 ± 8	726 ± 5	0.65	30.5 ± 0.2	22,143	227
1.2 Si	499 ±10	760 ± 8	0.66	26.8 ± 0.2	20,368	234

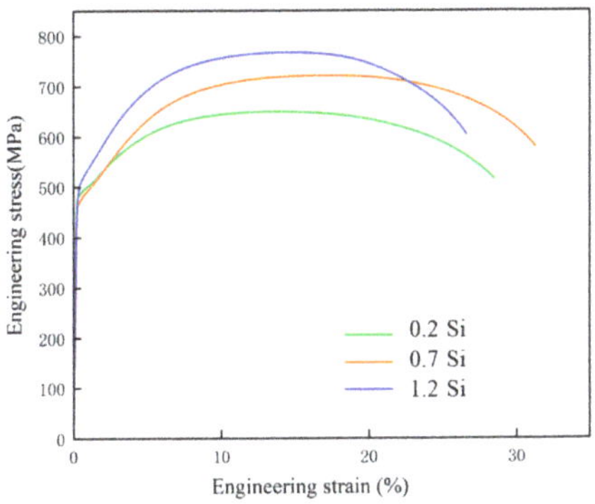

Figure 8. Stress–strain curve.

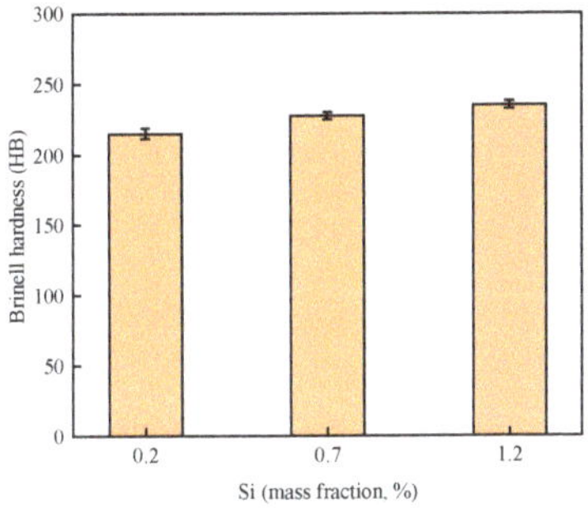

Figure 9. The Brinell hardness of the test steel with different Si contents under the tempering process.

The tensile strength of steel 0.2 Si is 652 MPa, the yield strength is 480 MPa, the elongation is 28.4%, the yield ratio is 0.74, and the strength–plasticity product reaches 18,517 MPa·%. The tensile strength of steel 0.7 Si is 726 MPa, the yield strength is 474 MPa, the elongation is 30.5%, the yield ratio is 0.65, and the strength–plasticity product reaches 22,143 MPa·%. The tensile strength of steel 1.2 Si is 760 MPa, the yield strength is 499 MPa, the elongation is 26.8%, the yield ratio is 0.66, and the strength–plasticity product reaches 20,368 MPa·%. As shown in Table 3, the tensile strength and hardness of the tested steel increase with the increase in Si content, while the elongation first increases and then decreases with the increase in Si content, and the addition of Si can reduce the yield ratio.

The continuous improvement in tensile strength is affected by two aspects. On the one hand, the solid solution of Si in ferrite plays a role in solid solution strengthening, which increases the strength and hardness of steel [27]. On the other hand, with the increase in Si content, the size of martensite in the test steel becomes finer, and the distribution becomes more uniform, which increases the effective grain boundary area of ferrite and martensite, increases the resistance to dislocation movement, and increases the shear stress required for deformation [28]. Therefore, the tensile strength and hardness of the test steel continue to improve. Generally speaking, the higher the strength and hardness of steel, the worse its elongation [29]. However, the elongation of the test steel first increases and then decreases because Si inhibits the formation of carbides, allowing carbon to diffuse from martensite to austenite. The enrichment of carbon in the original austenite is more complete, the stability is improved, and more retained austenite at room temperature is formed, which promotes the formation of retained austenite. During the plastic deformation process, due to the TRIP effect of retained austenite, the retained austenite undergoes a gradual martensitic transformation. This transformation process simultaneously enhances ductility, eliminates stress concentration, and delays the occurrence of necking, resulting in an increase in the elongation of the steel. Therefore, when the Si content is increased to 0.685 wt. %, the elongation increases. However, when the Si content is too high, a large amount of retained austenite grains are encapsulated within the soft ferrite matrix, resulting in the inability of retained austenite to exert its TRIP effect under stress and making it difficult to achieve an increase in plasticity. Therefore, when the Si content is increased to 1.206 wt. %, the elongation of the test steel decreases.

When the Si content increases to 0.685 wt. % and 1.206 wt. %, the yield ratio of the test steel decreases from 0.74 to 0.65 and 0.64, respectively. This is due to the increase in Si content, which clarifies the boundary between ferrite and martensite tempered structures. As a strengthening phase, martensite induces a large number of mobile dislocations within the adjacent ferrite, which can be activated at lower stress levels. Therefore, the test steel with Si contents of 0.685 wt. % and 1.206 wt. % exhibits low yield ratio characteristics. The strength–ductility product is a key index for evaluating the comprehensive mechanical properties of steel, which can directly reflect the forming and impact resistance of steel [27]. When the Si content is 0.685 wt. %, the steel has the best strength–ductility product, indicating that the steel has the best comprehensive performance.

3.4. The Fracture Morphology of the Test Steel after Tempering

Figure 10 shows the micro-morphology of the tensile sample fracture at different Si contents. There are unevenly sized circular or elliptical dimples in the micro-morphology of each sample, indicating that each sample has a slow tearing process, which will consume a large amount of plastic deformation work during the deformation process. The samples with larger size and deeper dimples have better plasticity. As shown in Figure 10a,b, the micro-morphology of the fracture of the samples with Si contents of 0.18 wt. % and 0.685 wt. % shows large and deep dimples, and the fracture is dominated by dimples. Combined with Table 3, the elongation is higher, and the plasticity is better. However, the micro-morphology of the fracture of the sample with a Si content of 1.206 wt. % shows relatively few dimples, with more cleavage planes and tearing edges, as shown in Figure 10c, which indicates poor elongation.

Figure 10. Fracture morphology of different Si content test steel after heat treatment: (**a**) 0.2 Si; (**b**) 0.7 Si; (**c**) 1.2 Si.

4. Conclusions

1. The microstructure of the test steel after quenching at 900 °C and intercritical quenching at 720 °C is mainly composed of ferrite and martensite structures. As the Si content increases, the ferrite content in the microstructure decreases, and the ferrite and martensite structures become finer and more evenly distributed.
2. After quenching at 900 °C, intercritical quenching at 720 °C, and intercritical tempering at 680 °C, the microstructure of the steel is mainly composed of a dual-phase structure of ferrite and tempered martensite, with a small amount of retained austenite and cementite. With the increase in Si content, the boundary between ferrite and martensite becomes clear after tempering, and the structure of tempered lath martensite becomes clear gradually.
3. The increase in Si content gradually increases the tensile strength and hardness of the test steel, with a slight decrease in the yield ratio and an initial increase followed by a decrease in elongation. When the Si content is 0.685 wt. %, the comprehensive performance of the test steel is optimal, with a tensile strength of 726 MPa, a yield ratio of only 0.65, the highest elongation of 30.5%, and the highest strength–ductility product of 22,143 MPa·%.

Author Contributions: Z.H.: experimental design, data collection, data analysis, and manuscript writing; H.F.: first draft revision, project management, technical supervision, and manuscript writing. All authors have read and agreed to the published version of the manuscript.

Funding: The authors would like to thank the financial support for this work from Hebei Science and Technology Major Project (22281005Z).

Data Availability Statement: The raw data supporting the conclusions of this article will be made available by the authors on request.

Conflicts of Interest: The authors declare no conflict of interest.

ferences

Weng, Y.Q.; Yang, C.F.; Shang, C.J. Development status and trend of low alloy steel in China. *Steel* **2011**, *46*, 1–10.
Qiao, Y.; Wang, G. Recent Status of Production, Administration Policies, and Low-Carbon Technology Development of China's Steel Industry. *Metals* **2024**, *14*, 480. [CrossRef]
Gao, G.; Liu, R.; Wang, K.; Gui, X.; Misra, R.D.K.; Bai, B. Role of retained austenite with different morphologies on sub-surface fatigue crack initiation in advanced bainitic steels. *Scr. Mater.* **2020**, *184*, 12–18. [CrossRef]
Xiao, H.; Zhao, G.; Xu, D.; Cheng, Y.; Bao, S. Effect of Microstructure Morphology of Q&P Steel on Carbon and Manganese Partitioning and Stability of Retained Austenite. *Metals* **2022**, *12*, 1613. [CrossRef]
Wu, Z.X.; Chen, X.; Liu, Y.; Li, Y.X.; Zhang, H.W. Effect of intercritical tempering treatment on the microstructure and mechanical properties of high-alloy cast steel. *Foundry* **2022**, *70*, 1156–1161.
Miller, R.L. Ultrafine-grained microstructures and mechanical properties of alloy steels. *Met. Trans.* **1972**, *3*, 905–912. [CrossRef]
Niikura, M.; Morris, J.W. Thermal processing of ferritic 5Mn steel for toughness at cryogenic temperatures. *Met. Trans. A* **1980**, *11*, 1531–1540. [CrossRef]
Lee, Y.-K.; Han, J. Current opinion in medium manganese steel. *Mater. Sci. Technol.* **2015**, *31*, 843–856. [CrossRef]
Xie, Z.-J.; Shang, C.-J.; Wang, X.-M.; Han, G.; Misra, R.-D. Recent progress in third-generation low alloy steels developed under M3 microstructure control. *Int. J. Miner. Met. Mater.* **2020**, *27*, 1–9. [CrossRef]
Xi, X.; Wang, J.; Chen, L.; Wang, Z. Tailoring Mechanical Properties of a Low Carbon Cu-Containing Structural Steel by Two-Step Interintercritical Heat Treatment. *Met. Mater. Int.* **2019**, *25*, 1477–1487. [CrossRef]
Huang, L.; Deng, W.T.; Liu, J.; Wang, Z. Relationship between retained austenite stability and low-temperature toughness in 0.12C-3.0Mn low-carbon medium-manganese steel. *J. Met.* **2017**, *53*, 316–324.
Li, D.J.; Bai, Q.; Chen, J.J.; Lv, N.; Feng, Y.R.; Ren, F.Z. Effect of intercritical heat treatment on the microstructure and mechanical properties of low-alloy TRIP steel. *J. Mater. Heat Treat.* **2019**, *40*, 111–117.
Liu, L.; He, B.; Huang, M. The role of transformation-induced plasticity in the development of advanced high strength steels. *Adv. Eng. Mater.* **2018**, *20*, 1701083. [CrossRef]
Sun, B.; Fazeli, F.; Scott, C.; Guo, B.; Aranas, C., Jr.; Chu, X.; Jahazi, M.; Yue, S. Microstructural characteristics and tensile behavior of medium manganese steels with different manganese additions. *Mater. Sci. Eng. A* **2018**, *729*, 496–507. [CrossRef]
Li, Z.C.; Misra, R.D.K.; Cai, Z.H.; Li, H.X.; Ding, H. Mechanical properties and deformation behavior in hot-rolled 0.2 C-1.5/3Al-8.5 Mn-Fe TRIP steel: The discontinuous TRIP effect. *Mater. Sci. Eng. A* **2016**, *673*, 63–72. [CrossRef]

16. Cai, Z.H.; Wang, S.K.; Zhou, Y.J.; Li, Z.C. The synergistic effect of grain refinement and precipitation strengthening on mechani properties and dry sliding wear behavior of medium manganese steels. *Tribol. Int.* **2023**, *179*, 108158. [CrossRef]
17. Zhang, B.G.; Zhang, X.M.; Liu, H.T. Microstructural evolution and mechanical properties of Ni-containing light-weight mediu Mn TRIP steel processed by intercritical annealing. *Mater. Sci. Eng. A* **2020**, *793*, 139289. [CrossRef]
18. Zhang, S.; Zhou, W.; Hu, F.; Yershov, S.; Wu, K. The role of Si in enhancing the stability of residual austenite and mechani properties of a medium carbon bainitic steel. *J. Mater. Res. Technol.* **2024**, *30*, 1939–1949. [CrossRef]
19. Kim, B.; Sietsma, J.; Santofimia, M.J. The role of silicon in carbon partitioning processes in martensite/austenite microstructu *Mater. Des.* **2017**, *127*, 336–345. [CrossRef]
20. Hwang, S.; Bai, Y.; Gao, S.; Park, M.-H.; Shibata, A.; Tsuji, N. Mechanism of simultaneous increase of strength and ductility w grain refinement in Si-added high-Mn austenitic steel. *Mater. Sci. Eng. A* **2024**, *894*, 146193. [CrossRef]
21. Song, C.; Zhang, Z.; Wu, W.; Wang, H.; Sun, Z.; Yang, Y.; He, W.; Xu, J.; Xia, Y.; Yin, W.; et al. Effect of Si on the dislocation st within martensite of ultra-high strength hot-rolled medium Mn steel with good ductility. *Mater. Sci. Eng. A* **2023**, *869*, 1448 [CrossRef]
22. Nouri, A.; Saghafian, H.; Kheirandish, S. Effects of Si Content and Interintercritical Annealing on Manganese Partitioning in D Phase Steels. *J. Iron Steel Res.* **2010**, *17*, 44–50. [CrossRef]
23. Zhu, Z.F.; Zhao, Z.Z.; Zhao, A.M.; Tang, D. Effect of Si on the microstructure and properties of high-strength hot-rolled dup steel. *Roll. Steel* **2011**, *28*, 16–18.
24. *GB/T 228.1-2021*; Metallic Materials-Tensile Testing—Part 1: Method of Test at Room Temperature. China Standards Publishi House: Beijing, China, 2021.
25. *GB/T 231.1-2018*; Metallic Materials-Brinell Hardness Test—Part 1: Test Method. China Standards Publishing House: Beiji China, 2018.
26. Xie, Z.J.; Yuan, S.F.; Zhou, W.H.; Yang, J.R. Stabilization of retained austenite by the two-step interintercritical heat treatment a its effect on the toughness of a low alloyed steel. *Mater. Des.* **2014**, *59*, 193–198. [CrossRef]
27. Xu, Y.; Cao, W.; Huang, M.; Zhang, F. First Principles Study of the Effects of Si, P, and S on the $\sum$5 (210)[001] Grain Boundary γ-Fe. *Metals* **2024**, *14*, 471. [CrossRef]
28. Shi, N.; Liu, W.M.; Liu, Y.G.; Zhan, H.; Cui, L.; Pan, H.B. Effects of Si content and quenching temperature on the microstruct and properties of Q&P steel. *Met. Heat Treat.* **2022**, *47*, 130–135. [CrossRef]
29. Shin, S.-H.; Oh, D.-K.; Hwang, B. Influence of Heat Treatment on Microstructure and Mechanical Properties of Direct-Quench Fe-0.06C-0.2Si-2.0Mn Steel. *Metals* **2023**, *13*, 1912. [CrossRef]

Disclaimer/Publisher's Note: The statements, opinions and data contained in all publications are solely those of the individu author(s) and contributor(s) and not of MDPI and/or the editor(s). MDPI and/or the editor(s) disclaim responsibility for any injury people or property resulting from any ideas, methods, instructions or products referred to in the content.

metals

icle

tudy on Fracture Behavior and Toughening Mechanisms of ltra-High-Strength Pipeline Steel

Li [1,2,3,*], Xiaoshun Zhou [1], Shujun Jia [2], Xiaoping Chen [2], Song Fu [3], Dongliang Zhao [3], Haonan Zhang [3] d Jie Guo [4]

1 College of Chemistry and Life Sciences, Zhejiang Normal University, Jinhua 321004, China; xszhou@zjnu.edu.cn
2 Central Iron and Steel Research Institute Company Limited, Beijing 100081, China; jiashujun@cisri.com.cn (S.J.)
3 Zhejiang Innuovo Magnetics Co., Ltd., Dongyang 322118, China
4 PetroChina Planning and Engineering Institute, Beijing 100083, China; guo_jie@petrochina.com.cn
* Correspondence: balicugb@sina.com

Abstract: In this paper, a series of low-temperature CVN (Charpy V-notch impact test) and DWTT (drop-weight tear test) experiments were carried out to deal with the intensifying contradiction of strength and toughness of ultra-high-strength pipeline steel. The fracture behavior and toughening mechanisms of ultra-high-strength pipeline steel were investigated using scanning electron microscopy, transmission electron microscopy and backscattered electron diffraction systems. The results show that DWTT fractures in ultra-high-strength pipeline steel had a variety of unconventional morphological features compared to CVN fractures, including ridge protrusion in ductile fracture conditions and a large-size fracture platform in brittle fracture conditions. Therefore, DWTT fractures contained more information about the material fracturing process, and could better reflect the actual process of material fracturing. In ultra-high-strength pipeline steel, fine-grained granular bainite caused cracks to undergo large deflections or frequent small transitions, which consumed additional energy and improved toughness. In contrast, large-sized granular bainite, which consisted of low-angle grain boundaries, did not effectively prevent crack propagation when it encountered cracks, which was not conducive to improved toughness. Moreover, the M/A constituents in large-sized granular bainite aggregated, cracked, or fell off, which could easily lead to the formation of microcracks and was also detrimental to toughening.

Keywords: ultra-high-strength pipeline steel; CVN; DWTT; toughening mechanism

1. Introduction

Increasing the strength of the pipeline steel can correspondingly increase the design pressure and thus the throughput. At the same time, improving the steel grade can also effectively reduce the wall thickness and reduces the difficulty of steel pipe manufacturing, which can not only better ensure the quality of the product, but also saves the amount of steel used and reduces the cost of steel pipe procurement [1–4].

However, as the strength of the metal materials continues to increase, the toughness decreases significantly, that is, there is an inversion between the strength and the toughness of the metal material. With the increase in the tensile strength of pipeline steel, the upper-platform energy and elongation decrease, and the tough–brittle transition temperature increases [5–7]. Arguably, strength is the foundation of the economic construction of pipelines, and toughness is the guarantee of safe pipeline service. This is one of the key issues that researchers are focusing on to improve both strength and toughness.

Generally, Charpy impact toughness is considered to be an essential index of dynamic ductile fracture arrest in gas pipelines. When the strength of pipeline steel exceeds grade X80, the crack arrest toughness index increases significantly, and the contradiction of strength and toughness intensifies. For ultra-high-strength pipeline steel, how to improve its fracture toughness to ensure the safe operation of the pipeline is an important direction of future research and application.

The toughening mechanism below the X100 level has been well studied. There, a good match of strength and toughness was achieved [8–14]. The contradiction between strengthening and toughening becomes more prominent when the steel grade reaches X100 [7,15–19]. However, research on the toughening mechanism of X100 ultra-high-strength pipe steel is highly insufficient.

In addition, for pipeline steel, especially for large-diameter ultra-high-strength pipeline steel, the results of the Charpy V-notch impact test (CVN) alone cannot completely evaluate the toughness of pipeline steel. The toughness level of a real steel pipe is affected by various factors such as the size, shape and stress state of the sample, and the geometrical constraints for small-size samples in CVN experiments are much smaller than those for actual components. Therefore, it is necessary to evaluate the toughness level of ultra-high-strength pipe steel more accurately using experimental methods that can better simulate real pipes. Among them, the Drop-Weight Tear Test (DWTT) is a relatively common large-scale experimental method in pipeline steel, which can give the relationship between the shear area of the fracture and the temperature. The results of DWTT experiments are in good agreement with those of the blasting test for pipelines [20–22].

Since DWTT samples are normally wall-thick pipes, the fracture surface contains all the information on the fracturing process, which plays an essential role in the deep understanding of crack propagation. However, the research on the DWTT of pipeline steel mainly focuses on the factors affecting its performance [23–29] and the analysis of abnormal fractures [30–33]. Some scholars have studied the morphological characteristics of some typical regions of DWTT fractures [34,35]. In contrast, there are few reports on the fine characterization and analysis of the DWTT fracture surface.

In this paper, we have investigated the fracture behavior of ultra-high-strength pipe steel by means of a series of temperature CVN experiments and a fine characterization and analysis of the low-temperature DWTT fracture surface, elucidating the mechanism of toughening and providing a theoretical reference for improving the toughness of ultra-high-strength pipe steels.

2. Experimental Material and Procedures

In this work, the experimental steel was produced by industrially hot-rolled steel plates. As shown in Figure 1a, the microstructure of ultra-high-strength pipeline steel was mainly a mixture of granular bainite (GB) and bainitic ferrite (BF). Then, the thermo-mechanical control process schedule is displayed in Figure 1b.

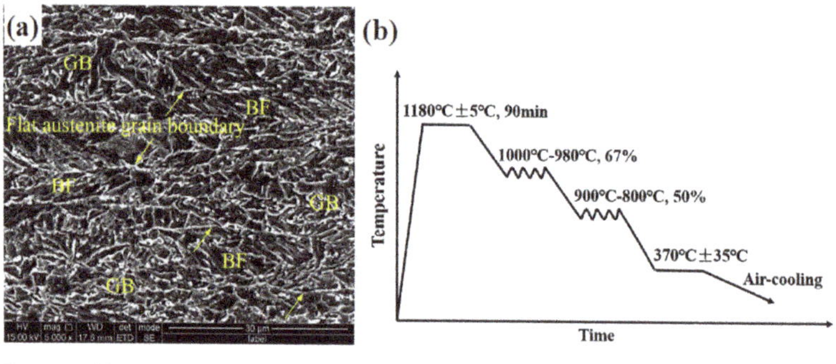

Figure 1. (**a**) Microstructure of ultra-high-strength pipeline steel; (**b**) Thermo-mechanical control process schedule.

The chemical composition and tensile properties of the experimental steel are listed in Tables 1 and 2, respectively. Two parallel specimens were processed for the tensile tests. It can be seen that the experimental steel reaches the X100 steel level of strength. X100 ultra-high-strength pipeline steel strips are usually formed by spiral welding. In this welding method, the weld is at a 30° angle to the rolling direction, and the mechanical properties in the 30° direction are the most significant. Therefore, the experimental specimens for the toughness study in this paper were also taken from the direction of 30° from the rolling direction of the steel plate.

Table 1. Chemical composition of experimental steel (wt. %).

C	Si	Mn	P	S	Nb	V	Ti	Mo	Ni	Cr	Cu
0.06	0.27	1.93	0.009	0.001	0.089	0.027	0.019	0.27	0.26	0.29	0.27

Table 2. Tensile properties of experimental steel at 30° direction.

Name	$R_{t0.5}$/MPa	R_m/MPa	$R_{t0.5}/R_m$	A/%
Specimen-1	700	805	0.87	28.5
Specimen-2	710	805	0.88	29

A series of CVN tests at 0 °C, −20 °C, −40 °C, −60 °C, −80 °C and −196 °C were performed on the impact test machine (JBN-300B, Wance Testing Machine Co., Ltd., Shenzhen, China) according to ASTM E23 [36]. The specimen size was 10 mm × 10 mm × 55 mm. Three parallel specimens were processed and their mean values were taken after the experiment. A low-temperature DWTT test was carried out on the instrumental drop-weight tear tester (DIT504C, Wance Testing Machine Co., Ltd., Shenzhen, China). The size of the DWTT specimen is referred to standard GB T 8363-2018 [37]. The experimental temperatures were −0 °C and −20 °C, and the size of the specimen was 305 mm × 75 mm × original thickness of the specimen (17.8 mm). Three parallel specimens were processed for the DWTT testing. Liquid nitrogen was used to control the experimental temperature so that the test was in the desired low-temperature state.

Then, the microstructure was represented by scanning electron microscopy (SEM, FEI Quanta 650FEG, FEI, Hillsboro, OR, USA), transmission electron microscopy (TEM, Tecnai G2F20, FEI, Hillsboro, OR, USA) and electron backscattered diffraction (EBSD, Oxford Nordlys F+, Oxford, UK). The SEM specimen was ground, polished, and etched in a 4% Nital solution. The TEM specimens were first mechanically ground to less than 50 µm and punched into a diameter of 3 mm, and then prepared with 6% perchloric acid and 94% methanol for twin-jet electrolytic polishing with the current of 50–80 mA and a temperature of −30 °C. The EBSD specimens were electropolished using an electrolyte consisting of 10% perchloric acid and 90% acetic acid.

3. Results and Discussion

3.1. *Fracture Behavior of Ultra-High-Strength Pipeline Steel*

3.1.1. Series of Low Temperature CVN Tests of Ultra-High-Strength Pipeline Steel

The low-temperature CVN results for the series are listed in Table 3. As can be seen from the table, the experimental steel satisfied the standard requirements of API Spec5L up to −60 °C. Then, starting from −40 °C, low values gradually appeared in parallel samples, and the impact energy decreased significantly at −80 °C.

Table 3. Impact energy of experimental steel at 30° direction (J).

Temperature	0 °C	−20 °C	−40 °C	−60 °C	−80 °C	−196 °C
1	310	315	292	253	69	5.9
2	299	300	291	260	74	4.2
3	310	298	246	271	43	3.6
Mean value	306	304	276	261	62	4.6

Figure 2 shows the impact fracture morphology of ultra-high-strength pipeline steel at low magnification. The fracture morphology from room temperature to −60 °C showed a large number of small deep dimples and large dimples interleaved. The size and distribution of the large dimples give rise to a distinct pattern. From the point of view of a microscopic mechanism, dimples are a fundamental feature of micropore aggregation and fracture as a result of micropore formation and growth. Larger and deeper micropores indicate that the material undergoes greater plastic deformation during fracture, consumes additional energy, and thus exhibits higher toughness [38]. Therefore, the morphology of the heterogenic dimple indicates a good toughness of the material. At −60 °C, microcracks were observed on the fracture marked by sky blue arrows. In the upper-right triangular frame region, the small dimples disappeared and only a few large dimples could be distinguished, and there were no features of cleavage fracture. Such morphological features indicate the onset of toughness reduction. When the experimental temperature was reduced to −80 °C, brittle fractures occurred in the experimental steel, and large cracks could be seen on the fracture marked by the sky-blue arrow. At −196 °C, the fracture morphology is also brittle fracturing, and the obvious tearing edge (marked by the yellow arrow) on the fracture can be identified as a quasi-cleavage fracture.

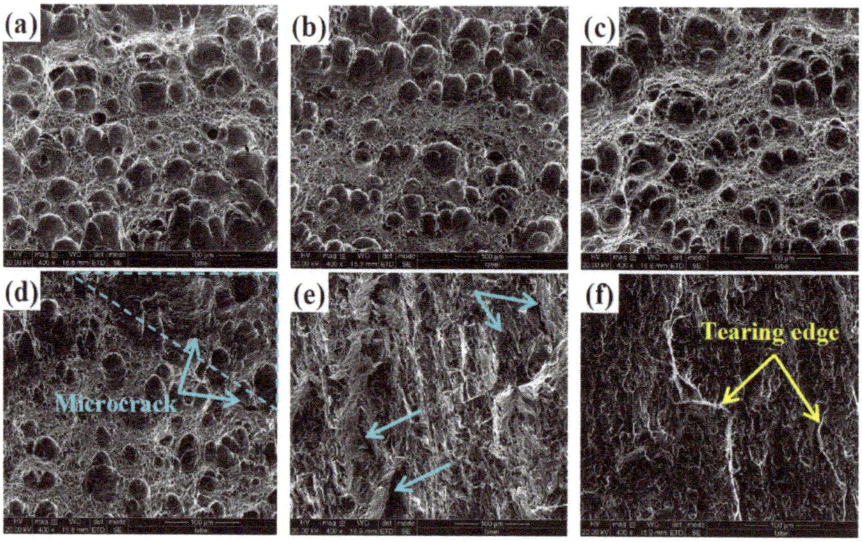

Figure 2. Low magnification SEM images of impact fracture morphology of ultra-high-strength pipeline steel: (**a**) 0 °C; (**b**) −20 °C; (**c**) −40 °C; (**d**) −60 °C; (**e**) −80 °C; (**f**) −196 °C.

At high magnification, a grooved crack could be observed at −80 °C, as shown by the sky-blue arrow in Figure 3a–d, which is usually caused by "delamination" during the fracture process [30–32]. A cleavage fracture occurs around the grooved crack in the plastic zone, and the red arrow in Figure 3b marks the dimple band distributed along the left edge of the grooved crack.

When the temperature was reduced to −196 °C, as shown in Figure 3e,f, it could be seen that there were tear edges distributed in different directions on the section (marked by yellow arrows), and the crack stopped propagating when it met the tear edges perpendicular to it (marked by sky-blue arrows), but would expand along the direction parallel to it.

Figure 4 shows the "displacement–load–energy" curve of ultra-high-strength pipeline steel at different temperatures. At the temperature of −60 °C and above, the crack-initiation energy is the same, and the total energy leading to the final fracture of the material is also similar, and the temperature change mainly affects the crack-arresting energy. Brittle fracture features appeared on the curve starting from −40 °C, as shown in the dotted box in Figure 4d,e, but ductile fracturing was still dominant on the whole. When the temperature was reduced to −80 °C, the curve changed significantly, as shown in Figure 4f, representing the disappearance of the plastic deformation region in the ductile fracture stage. This indicates that the fracture at this time transformed into a brittle fracture. In addition, the total energy consumed by the fracture also decreased sharply, indicating that the material failed rapidly and fractures occurred under brittle fracture conditions. This is consistent with the results observed for the fracture sections in Figures 2 and 3.

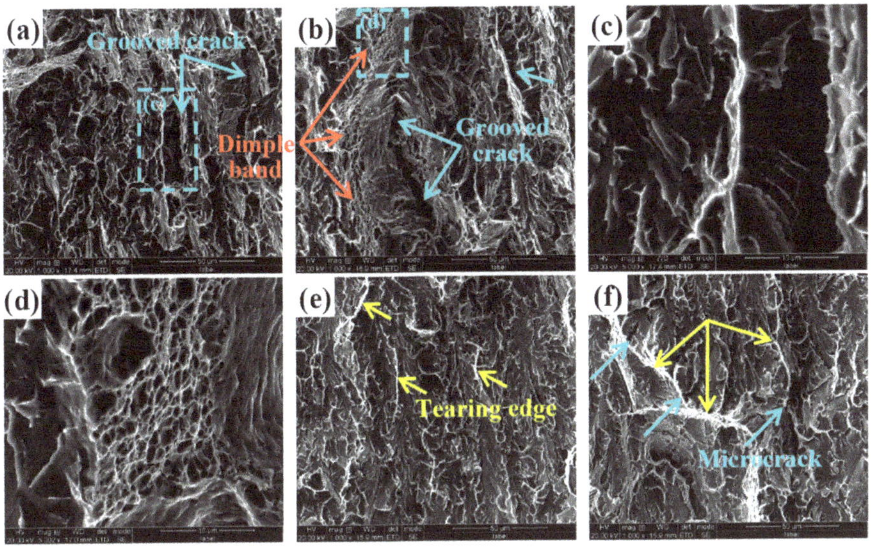

Figure 3. High-magnification SEM images of impact fracture morphology of ultra-high-strength pipeline steel: (**a**,**b**) −80 °C; (**c**) enlarged image of the sky-blue dotted box in (**a**); (**d**) enlarged image of the sky-blue dotted box in (**b**); (**e**,**f**) −196 °C.

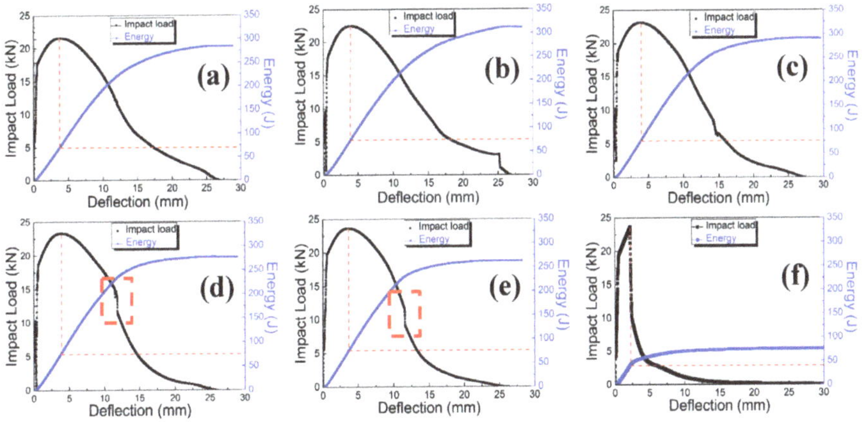

Figure 4. "Deflection–load–energy" curve of ultra-high-strength pipeline steel: (**a**) room temperature; (**b**) 0 °C; (**c**) −20 °C; (**d**) −40 °C; (**e**) −60 °C; (**f**) −80 °C.

3.1.2. Characterization and Analysis of DWTT Fracture of Ultra-High-Strength Pipeline Steel

Compared to the small-size samples in the CVN experiment, the large-size DWTT sample with full plate thickness is closer to the actual stress and strain state of the pipe at the time of fracture, which better reflects the long-term expansion of the crack in the actual pipe. At the same time, a significant increase in the section length can provide more information about the fracture morphology, which is conducive to an in-depth understanding of the fracturing process.

Figure 5 shows a macroscopic picture of a DWTT fracture in ultra-high-strength pipe steel. At 0 °C and −20 °C, the shear area of the DWTT fracture is 99% and 71.6%, respectively. That is, 85% of the standard index requirements cannot be met when the temperature is reduced to −20 °C.

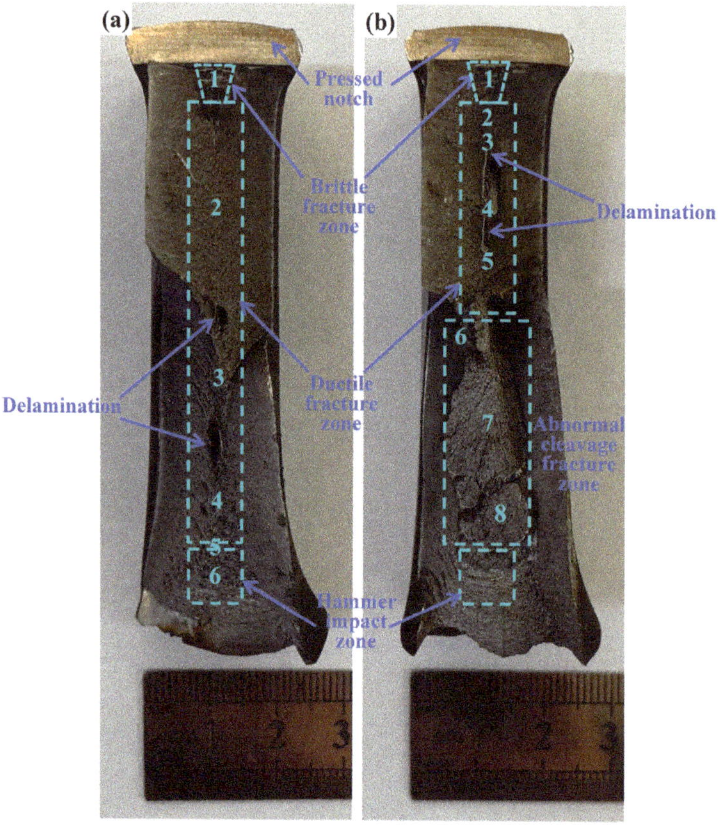

Figure 5. Macrograph of DWTT sample fracture of ultra-high-strength pipeline steel: (**a**) 0 °C; (**b**) −20 °C. Numbers 1–8 indicate the location of the observations.

DWTT fractures can usually be divided into four regions: the pressing notch, brittle fracture zone, ductile fracture zone, and hammer fracture zone. Among them, the brittle and ductile fracture zones are key analysis areas that can give various information about the fracturing process.

In particular, there are delamination defects on the surface of the fracture at both temperatures, and the delamination defect is a typical morphology when the cleavage plane (100) is parallel to the rolled surface during the fracturing process.

Texture analysis in the 30° direction of ultra-high-strength pipeline steel shows that there is a {001}<100> texture in the core structure, which is prone to stratification defects when the {001} texture is distributed along the elongation direction of the flat grain [21,39].

Figure 6 shows the microstructure morphology of position 1 in the brittle fracture zone of a DWTT fracture of ultra-high-strength pipeline steel at 0 °C. Although it had brittle-fracture characteristics in the macroscopic morphology, the SEM morphology at low magnification (Figure 6a) showed a large number of dimples with mixed sizes and some flat plane regions. When the planar region was enlarged (Figure 6b,c), it could be found that the dimples had totally disappeared, leaving only some irregular holes. The dimple disappears because a brittle fracture occurs at this location, that is, the fracture pattern of the material changes from ductile to quasi-cleavage. As a result, the dimple morphology typical of ductile fractures disappears at the fracture. Although it does not have the typical characteristics of a cleavage fracture, its contribution to toughness is extremely limited. Tear edges, which are common in quasi-cleavage fracture morphology, were observed as indicated by the yellow arrows in Figure 6d. In addition, a step-like morphological feature was also observed in some regions, as indicated by the blue arrows in Figure 6e,f, which is also one of the characteristics of quasi-cleavage fracture [40].

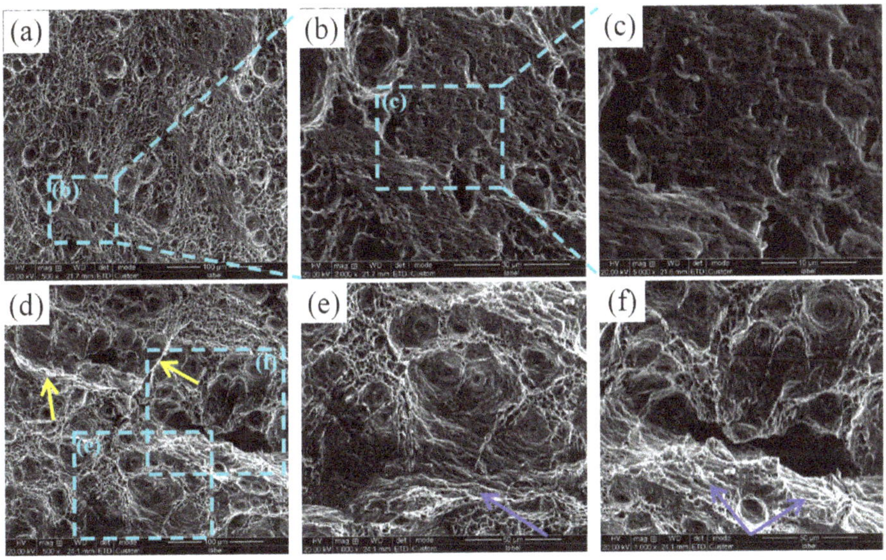

Figure 6. Micrograph of position 1 in DWTT fracture of ultra-high-strength pipeline steel at 0 °C. Figures (**a–f**) show different regions at position 1, respectively.

Figure 7 shows the microstructure morphology of position 2 in the ductile fracture zone of a DWTT fracture of ultra-high-strength pipeline steel at 0 °C. The fracture morphology was typical ductile fracture morphology with an alternating distribution of small and deep and large-size dimples, showing good toughness, as shown in Figure 7a–c. On the one hand, the formation of dimples is related to the size and distribution of the second-phase particles inside the microstructure, and on the other hand, it is also related to the grain size. The presence of fine granular bainite in ultra-high-strength steel favors the formation of small and deep dimples. In particular, ridge-like protrusion similar to the tear edge of quasi-cleavage features were observed at low magnification, as shown by the yellow arrows in Figure 7d. At high magnification (Figure 7e,f), it could be found that this special morphology was formed by severe plastic deformation of a large number of small dimples. In general, the formation of tear edges consumes a large amount of energy, which increases the impact energy. Microstructures similar to tear edges formed in the ductile fracture zones are not observed in the fracture morphology of the impact specimens.

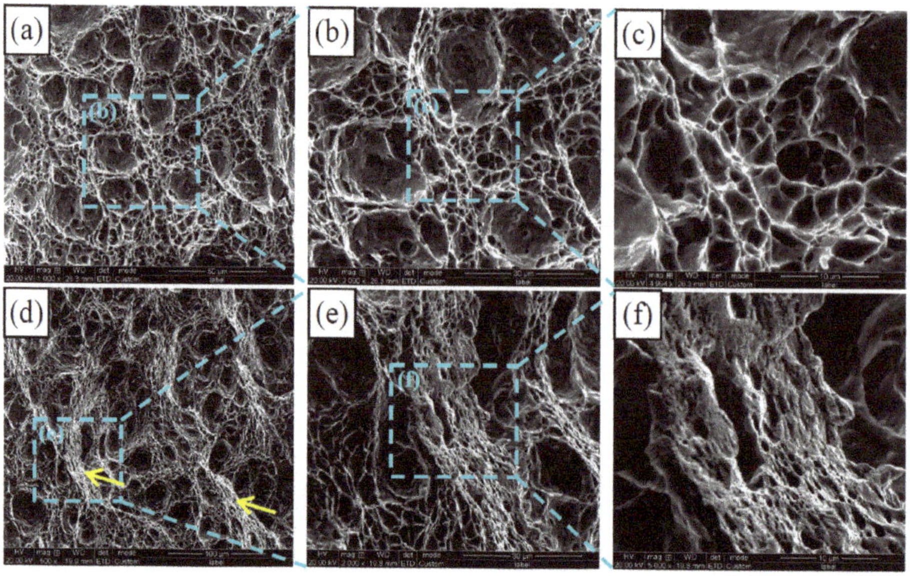

Figure 7. Micrograph of position 2 in DWTT fracture of ultra-high-strength pipeline steel at 0 °C. Figures (**a**–**f**) show different regions at position 2, respectively.

Figure 8 shows the microstructure morphology of position 3 in the ductile fracture zone of a DWTT fracture of ultra-high-strength pipeline steel at 0 °C. A ridge-like protrusion similar to that at position 2 is also observed at different magnifications, marked by the yellow arrow. Figure 9 shows the microstructural morphology of the ductile fracture zone at different magnifications of location 4. A sharper ridge-like protrusion is also observed, marked by a yellow arrow.

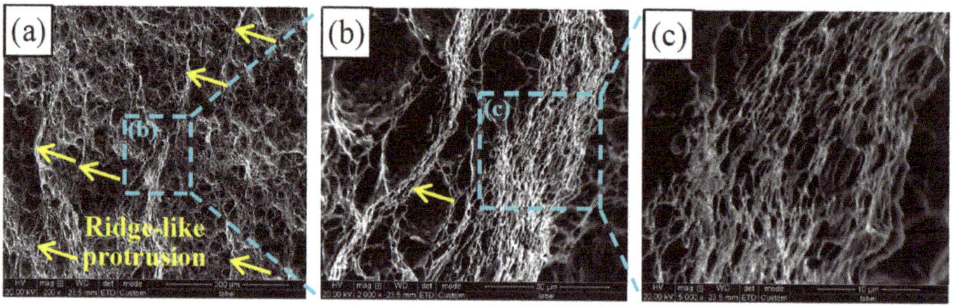

Figure 8. Micrograph of position 3 in DWTT fracture of ultra-high-strength pipeline steel at 0 °C. Figures (**a**–**c**) show different magnifications at position 3, respectively.

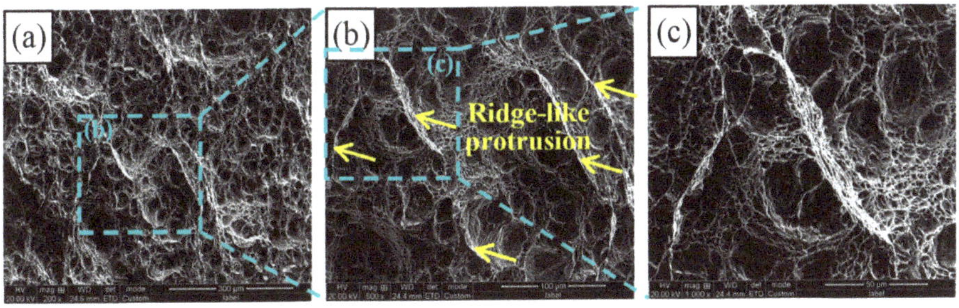

Figure 9. Micrograph of position 4 in DWTT fracture of ultra-high-strength pipeline steel at 0 °C. Figures (**a**–**c**) show different magnifications at position 4, respectively.

Although the macroscopic morphology of the fracture in Figure 5a shows the characteristics of a brittle fracture, the microscopic morphology remains ductile and also contains the tear-like edge structures observed at positions 2 and 3.

At position 5, the river pattern accompanied by micro-cracks (marked by sky-blue arrows) was observed, which is a typical brittle-fracture feature, as shown in Figure 10. However, position 5 has already reached the hammering zone and is therefore not included in the measurement range.

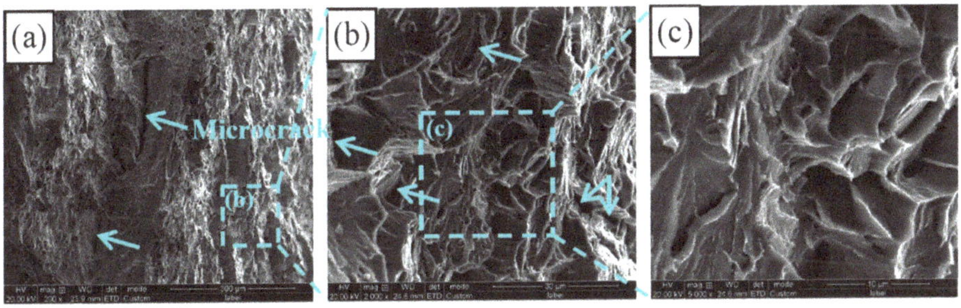

Figure 10. Micrograph of position 5 in DWTT fracture of ultra-high-strength pipeline steel at 0 °C. Figures (**a**–**c**) show different magnifications at position 5, respectively.

Figure 11 shows the microstructure morphology at position 6 of the hammering zone at different magnifications. Similarly to position 4, although the fracture morphology in Figure 8a is characterized by brittle fracturing at the macroscopic level, it still consists of ductile-fracture morphology at the microscopic level, as shown in Figure 11a–c. However, some of the dimples had significant plastic deformation (Figure 11d,e), or even had no dimple appearance at all (Figure 11f).

Observations of the microscopic morphology at positions 4 to 6 show that, unlike the fracture of the impact specimen, it is not correct to judge the fracture type by looking at the macroscopic morphological features of the DWTT specimen alone.

Figure 12 shows the microstructure morphology of position 1 in the brittle fracture zone of a DWTT fracture of ultra-high-strength pipeline steel at −20 °C. At low magnification, the fracture was extremely flat (Figure 12a,b), with bright white stripes on the surface (marked by yellow arrows in Figure 12b,c), accompanied by numerous microcracks (marked by blue arrows in Figure 12d,e). The overall morphology was characterized by brittle fracturing, but it differed from the river pattern in a typical cleavage fracture. In addition, an irregularly arranged step shape was observed around the crack, as shown in Figure 12f.

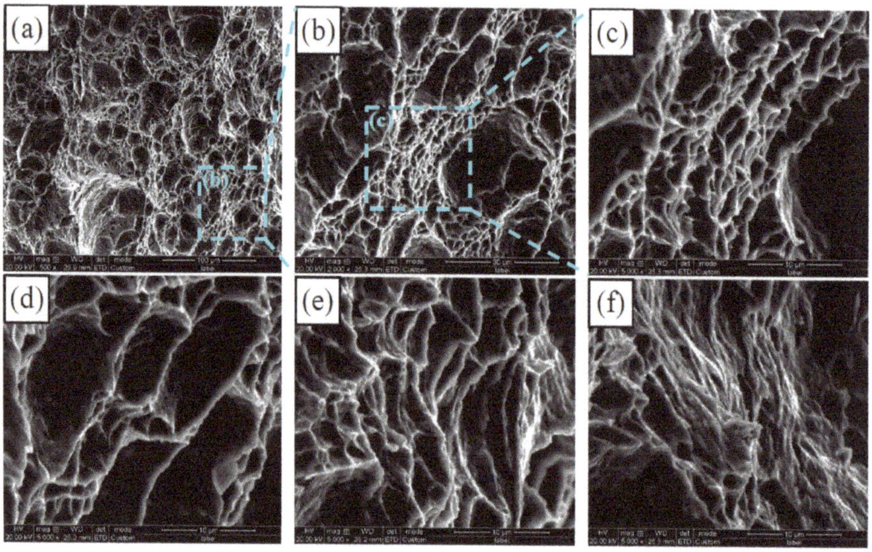

Figure 11. Micrograph of position 6 in DWTT fracture of ultra-high-strength pipeline steel at 0 °C. Figures (**a**–**f**) show different regions at position 6, respectively.

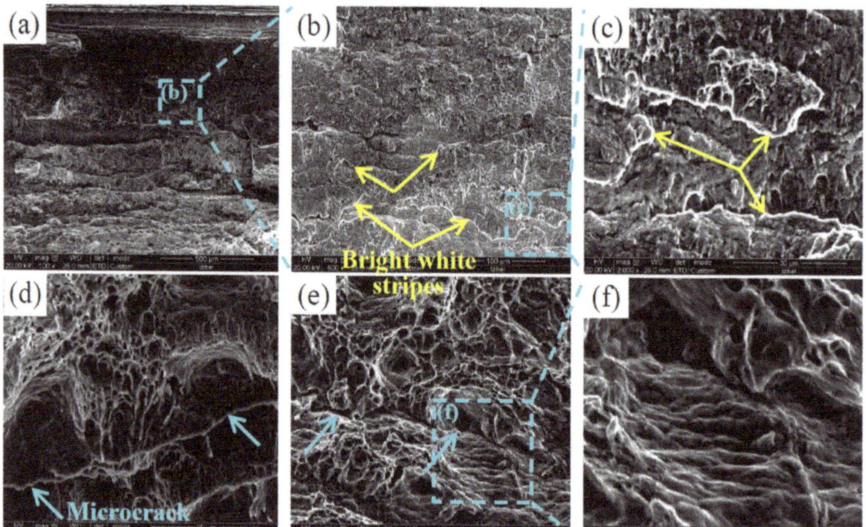

Figure 12. Micrograph of position 1 in DWTT fracture of ultra-high-strength pipeline steel at −20 °C. Figures (**a**–**f**) show different regions at position 1, respectively.

Figure 13 shows the microscopic morphology of position 2. At low magnification, the surface of the fracture is uneven (Figure 13a), and after local magnification, it shows a typical deformed dimple appearance (Figure 13b). Some of the dimples are shallow and small, with irregular morphology (Figure 13c). In some areas, a "ridge" shape consisting of a large number of distorted and elongated small dimples can also be observed (Figure 13e,f), which indicates that large plastic deformation and large energy consumption were experienced during the fracture process.

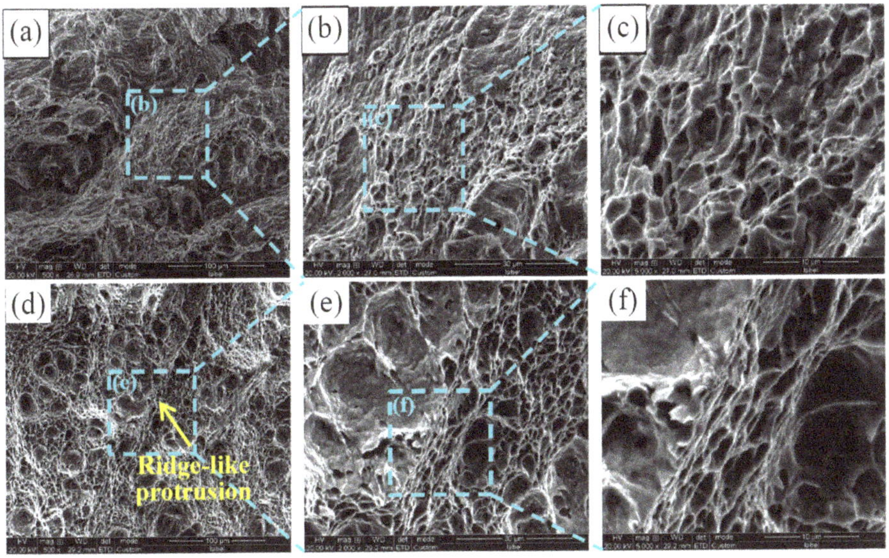

Figure 13. Micrograph of position 2 in DWTT fracture of ultra-high-strength pipeline steel at −20 °C. Figures (**a–f**) show different regions and magnifications at position 2, respectively.

Figure 14 shows the microscopic morphology of position 3 in the ductile fracture zone of a DWTT fracture at −20 °C. As can be seen from the macroscopic morphology of Figure 5b, delamination defects appeared at this position. More "ridge"-like structures similar to those in Figure 13 could be observed above the delamination defects, as shown in Figure 14a–d. The cracks of the delamination defects are surrounded by a large number of small dimples, as can be seen from the sides of the delamination defects in Figure 14e,f.

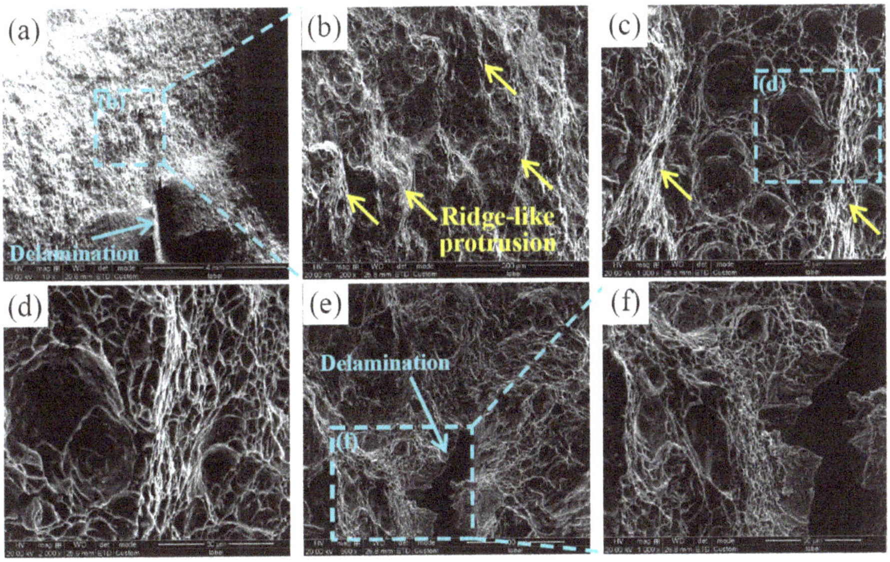

Figure 14. Micrograph of position 3 in DWTT fracture of ultra-high-strength pipeline steel at −20 °C. Figures (**a–f**) show different regions at position 3, respectively.

Figures 15 and 16 show the microstructures of position 4 and 5 in the DWTT fracture ductile fracture zone of X100 ultra-high-strength pipeline steel at −20 °C, respectively. As can be seen in Figure 15, the micromorphology at position 4 shows typical brittle cleavage-fracture features, including cleavage steps and river patterns. This indicates that local embrittlement occurred at the front end of the delamination defect. At position 5 under delamination defects, although a large number of edges parallel to the fracture direction can be observed at low magnification (Figure 16a), as at position 3, the entire section at high magnification shows deformed dimple morphology (Figure 16b,c), that is, it is still a ductile fracture. According to the results in Figures 14–16, the occurrence of delamination defects was independent of the fracture pattern.

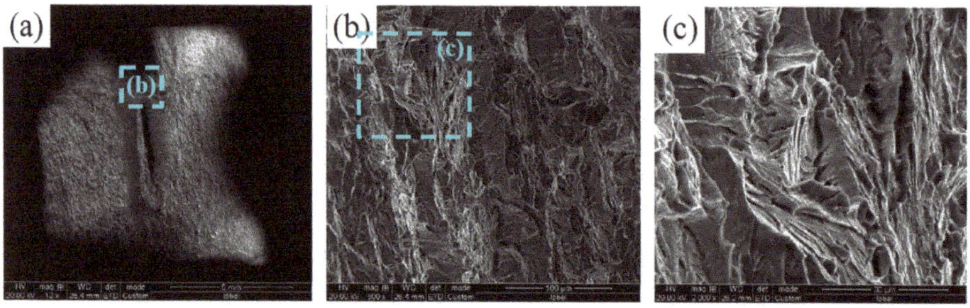

Figure 15. Micrograph of position 4 in DWTT fracture of ultra-high-strength pipeline steel at −20 °C. Figures (**a–c**) show different magnifications at position 4, respectively.

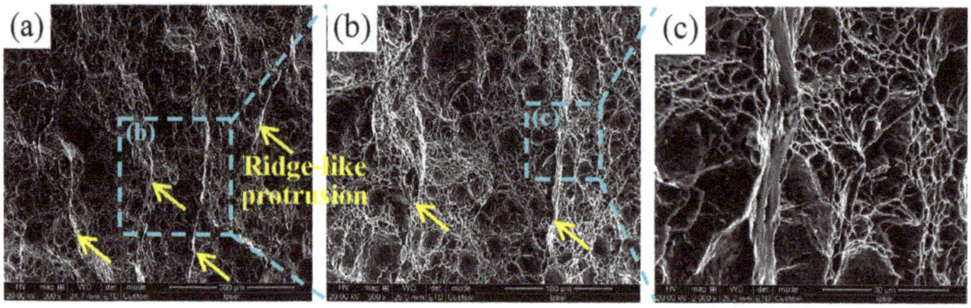

Figure 16. Micrograph of position 5 in DWTT fracture of ultra-high-strength pipeline steel at −20 °C. Figures (**a–c**) show different magnifications at position 5, respectively.

The micromorphology of ultra-high-strength pipeline steel at positions 6 to 8 within the brittle fracture zone of a DWTT fracture at −20 °C is shown in Figures 17–19, and the micromorphology at different positions in this region is all brittle fracturing. In general, cleavage fracturing is not carried out on a single cleavage plane, but along a set of parallel crystal faces, with cleavage planes at different heights connected by cleavage steps [41]. The blue arrows in Figure 17b,c mark the large cleavage steps, while the trend of the river pattern shows the direction of crack propagation. A large number of deforming dimples (marked by the red arrow) can be observed around the cleavage plane (marked by the blue arrow) in Figure 17e, indicating that the overall fracture pattern is a quasi-cleavage fracture. Position 7 is radial in the macroscopic morphology of Figure 5b and remains as a cleavage fracture in the microscopic morphology of Figure 18. There are a number of secondary microcracks marked by sky-blue arrows along the propagation direction of the crack. In particular, a large number of small dimples were observed on the side of a large cleavage step, as shown in Figure 18e,f, indicating that local plastic deformation also occurred during the brittle fracturing.

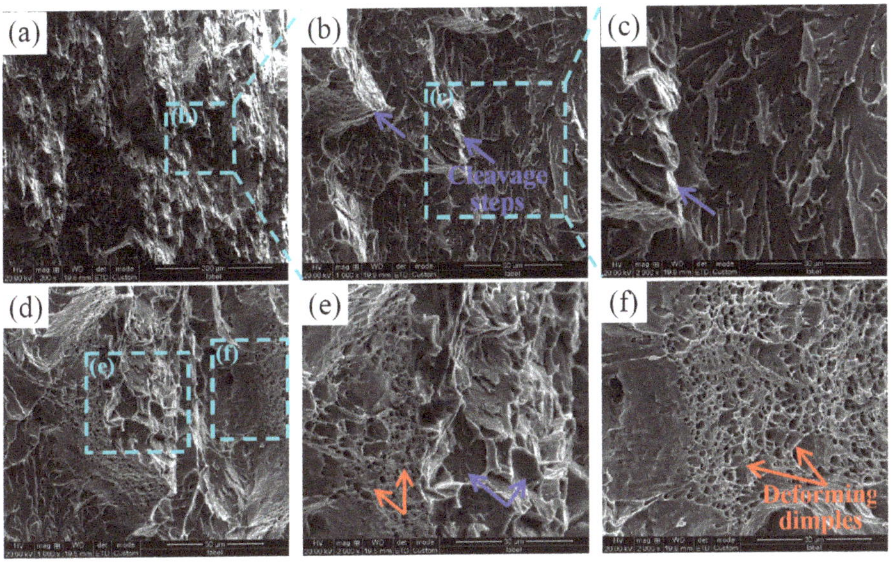

Figure 17. Micrograph of position 6 in DWTT fracture of ultra-high-strength pipeline steel at −20 °C. Figures (**a**–**f**) show different regions at position 6, respectively.

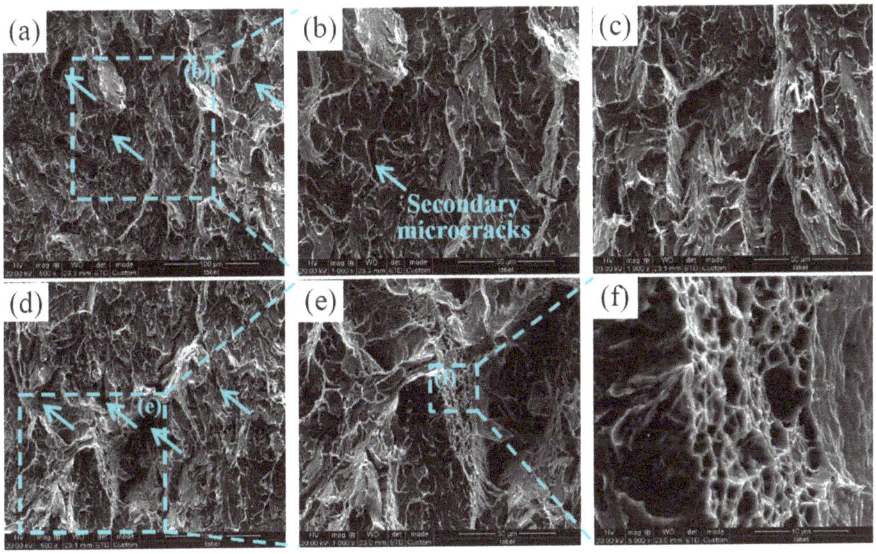

Figure 18. Micrograph of position 7 in DWTT fracture of ultra-high-strength pipeline steel at −20 °C. Figures (**a**–**f**) show different regions at position 7, respectively.

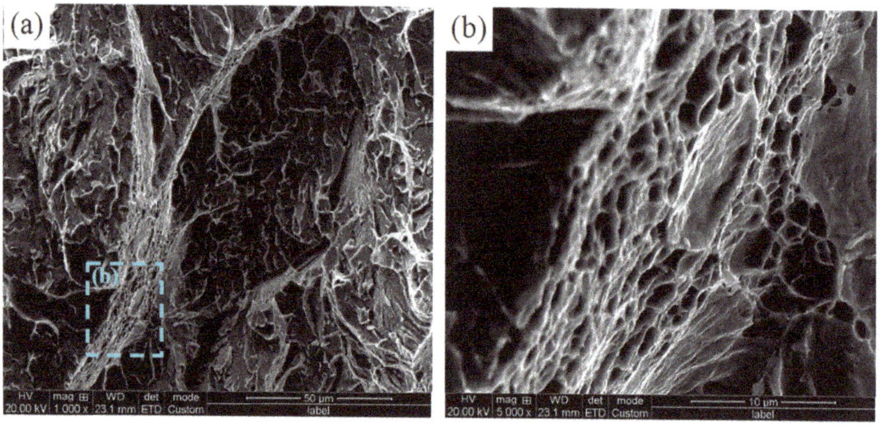

Figure 19. Micrograph of position 8 in DWTT fracture of ultra-high-strength pipeline steel at −20 °C. Figures (**a**,**b**) show different magnifications at position 8, respectively.

At position 8, a large secondary crack forms along the junction of the cleavage steps, as shown in Figure 19. Similar to Figure 18f, one side of the large-size cleavage step has a large number of small dimples that are deformed and elongated. The study of L. Cho et al. [40] pointed out that the formation of ridge-like and large-size step-like microstructures is related to large-angle grain boundaries or lath boundaries, which has a significant hindering effect on the propagation of cracks.

Detailed fracture observations of DWTT specimens indicate that in actual fracturing processes, even if a brittle fracture occurs, local plastic deformation will still accompany the crack propagation.

3.2. *Toughening Mechanism of Ultra-High-Strength Pipeline Steel*

3.2.1. Effect of Deforming Austenite Size on DWTT Performance

Figure 20 shows the parent austenite grains of X100 ultra-high-strength pipe steel in the thickness direction, showing the flat austenite grain boundary morphology after TMCP. The average flat austenite height in the overall thickness direction is about 9 μm. Relevant studies have pointed out that when the flattening height of austenite is less than 9 μm, it can have excellent DWTT performance [9,41]. Thus, the complete flattening of the austenite during the controlled rolling process is one of the reasons for ensuring the toughness of the X100 ultra-high-strength pipe steel.

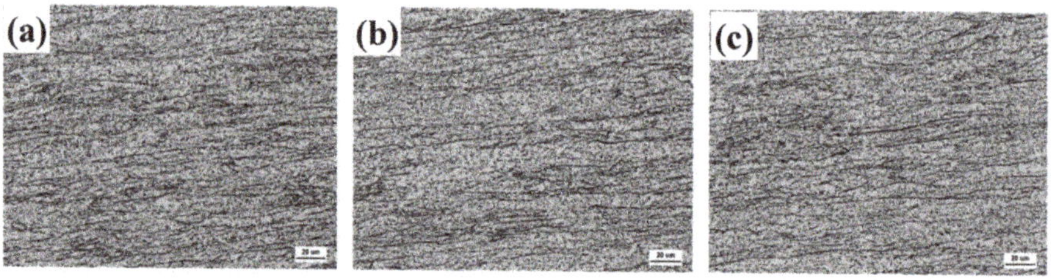

Figure 20. Metallograph of parent austenite grains of ultra-high-strength pipe steel in thickness direction: (**a**) side; (**b**) 1/4 position; (**c**) center.

3.2.2. Characterization and Analysis of Crack Propagation Path in Fracture Section

Figure 21 shows the CVN fracture section images of X100 ultra-high-strength pipeline steel at −80 °C and −196 °C. It can be seen that, when the crack propagation encounters the fine-grained granular bainite, large deflections or frequent small turns will occur, which will consume more energy and improve the toughness. Clustered grains in a fine-grained granular bainite will disperse the high stress at the crack tip to the neighboring grains through coordinated deformation, so that the crack will deflect or even stop propagating. In contrast, when the crack propagation encounters a large-sized granular bainite, it will pass directly through it, that is, the large-sized granular bainite lacks an effective blocking effect on the crack and thus has a limited contribution to the toughness. Typical microstructure morphologies of the two granular bainites are shown in Figure 22.

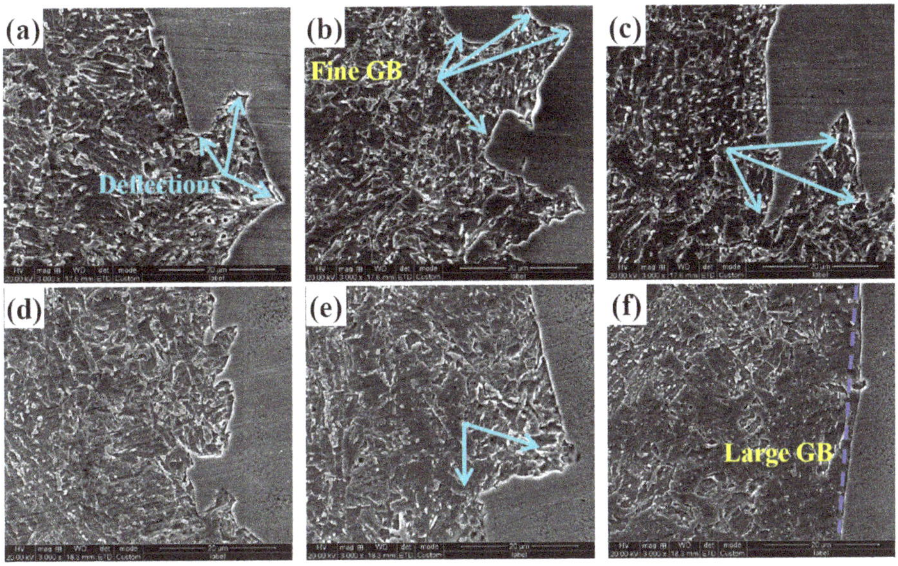

Figure 21. Fracture section images of CVN specimen of ultra-high-strength pipeline steel: (**a**–**c**) −80 °C; (**d**–**f**) −196 °C.

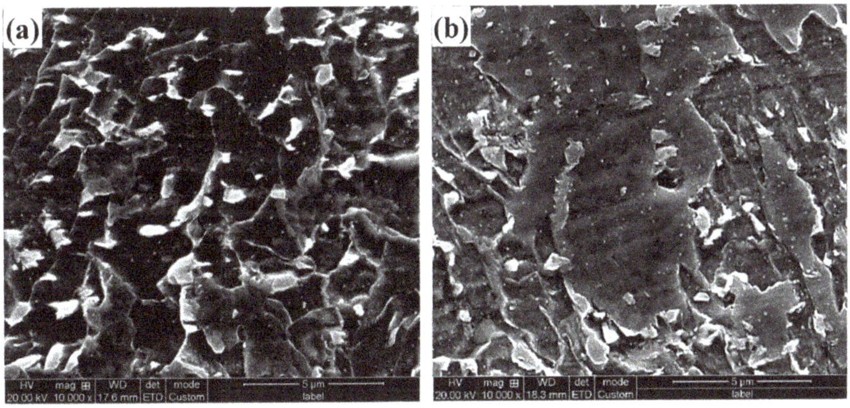

Figure 22. Fracture section microstructures of CVN specimen of ultra-high-strength pipeline steel: (**a**) fine-grained granular bainite; (**b**) large-sized granular bainite.

In order to further study the crack propagation process, EBSD pictures of the undeformed region of the fracture section (Figure 23a,b) and the cross-section of the crack propagation path (Figure 23c,d) are presented, respectively. It can be seen from Figure 23a,b that the large-size granular bainite was composed of a large number of low-angle grain boundaries (LAGB, interfaces where the misorientation between adjacent grains is 2°–15°). In contrast, the fine-grained bainite was surrounded by high angle grain boundaries (HAGB, the interface where the misorientation between adjacent grains is greater than 15°), and the internal substructures were few and distributed in clusters. As can be seen from Figure 23c,d, when the crack encountered these small grains in the propagation process, it would have a large transition and change the direction of propagation (marked by dashed and broken lines in Figure 23d). However, when it encountered the large-sized granular bainite, it passed directly through, in agreement with SEM observations. Large-sized granular bainite, which is mainly composed of LAGBs, cannot effectively block crack propagation when it encounters a crack, which is not conducive to toughness enhancement, as indicated by the red arrows in Figure 23d.

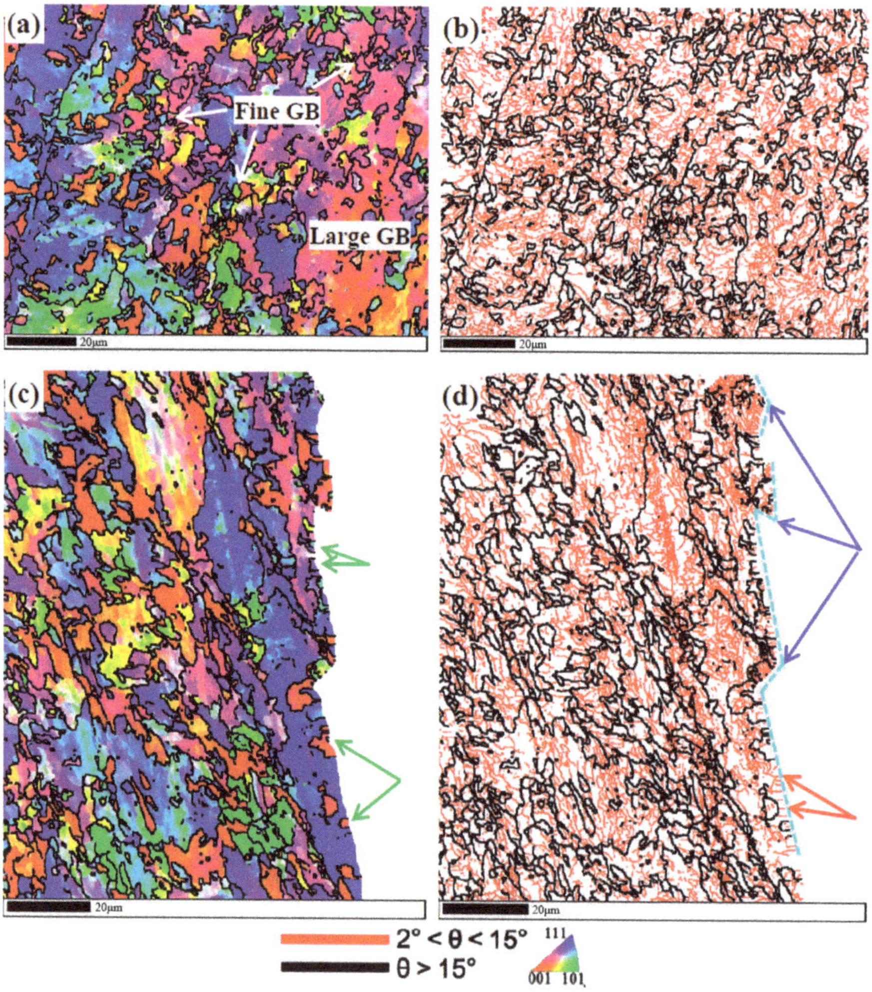

Figure 23. EBSD images of fracture section at −196 °C conditions: (**a**,**c**) inverse pole figure; (**b**,**d**) grain boundary figure. Green arrows show that the HAGBs did not deflect the crack. Blue arrows show that high-density substructures proceeded large transitions. Red arrows show that LAGBs could not effectively block crack propagation.

It is commonly believed that HAGBs may inhibit crack propagation. However, it could also be found in Figure 23c that, although some of the small grains distributed separately were surrounded by HAGBs, they did not deflect the crack, as indicated by the green arrows. Relevant studies have shown that not all HAGBs have an obstruction effect on crack propagation, and only the interface with an angle of ≥35° between adjacent grain {100} cleavage planes in pipeline steel have an obvious obstruction effect on crack propagation [41]. In contrast, the clumping of fine grains will disperse the high stress at the crack tip to neighboring grains through coordinated deformation, thus deflecting or even preventing crack propagation.

In Figure 23d, high-density substructures can be observed to form inside some of the small grains at several locations where large transitions proceeded, marked by blue arrows, indicating that coordinated deformation occurred when the crack propagation encountered clusters of fine grains. Grain refinement can increase the density of boundaries with a higher special misorientation angle between {110}-planes, which is conducive to suppressing crack initiation [8]. The majority of the dislocations tend to pile along the {110} slip planes as dense dislocation walls to coordinate deformation. At low strains, grains are first divided into cell blocks that are nearly dislocation-free. At higher strains, dislocation tangled zones and lamellar boundaries develop [42]. The cell blocks, dislocation tangled zones and lamellar boundaries all belong to the substructures as marked by blue arrows in Figure 23d. Furthermore, combined with the morphological analysis of the brittle fracture, it can be seen that the large-size step morphology on the brittle zone of the fracture was most likely formed by the giant transition during the propagation of the crack.

3.2.3. Analysis of Secondary Crack in Fracture Section

Figure 24 shows the morphology of secondary cracks in the brittle-fracture cross-section of X100 ultra-high-strength pipeline steel. It can be seen that the microstructure contained a large number of M/A constituents, which were mainly granular and striated. It could also be observed that some of the M/A constituents aggregated and led to crack formation, as shown by the dashed ellipse in Figure 24a. The M/A constituent belongs to the hard and brittle phase in the microstructure of pipeline steel and is generally surrounded by a stress field. If M/A constituents aggregate, the superposition of adjacent stress fields will produce stress concentration, and it is easy to crack and form micro-cracks in the process of deformation [43,44]. These cracks occur with a large deflection in the fine-grained granular bainite region, but pass directly through the large-sized granular bainite.

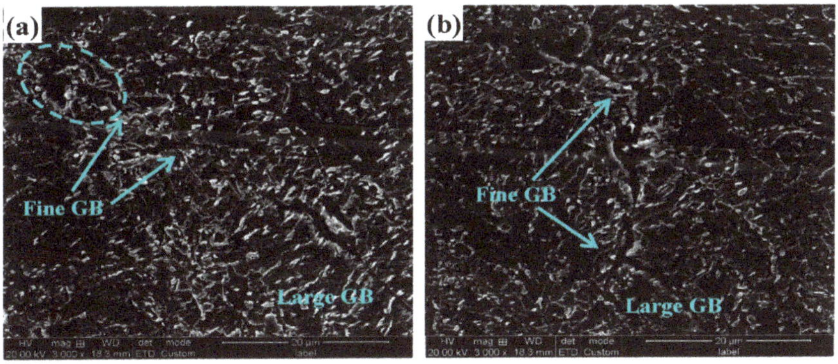

Figure 24. Microstructure of secondary cracks in fracture cross-section of CVN specimen at −196 °C conditions. Figures (**a**,**b**) show different regions at fracture cross-section.

Figure 25 further shows a high-magnification picture of the secondary crack in the cross-section of the −196 °C impact fracture. When the crack propagation encountered bainite ferrite in ultra-high-strength steel, it could pass directly through the bainite lath, but would deflect or stop at the sheaf boundary (indicated by the yellow dashed line in Figure 25a) or at the granular bainite grain boundary (indicated by the dashed ellipse in Figure 25b). In bainite ferrite, the interface of the lath is LAGB, so there is no blocking effect on the crack.

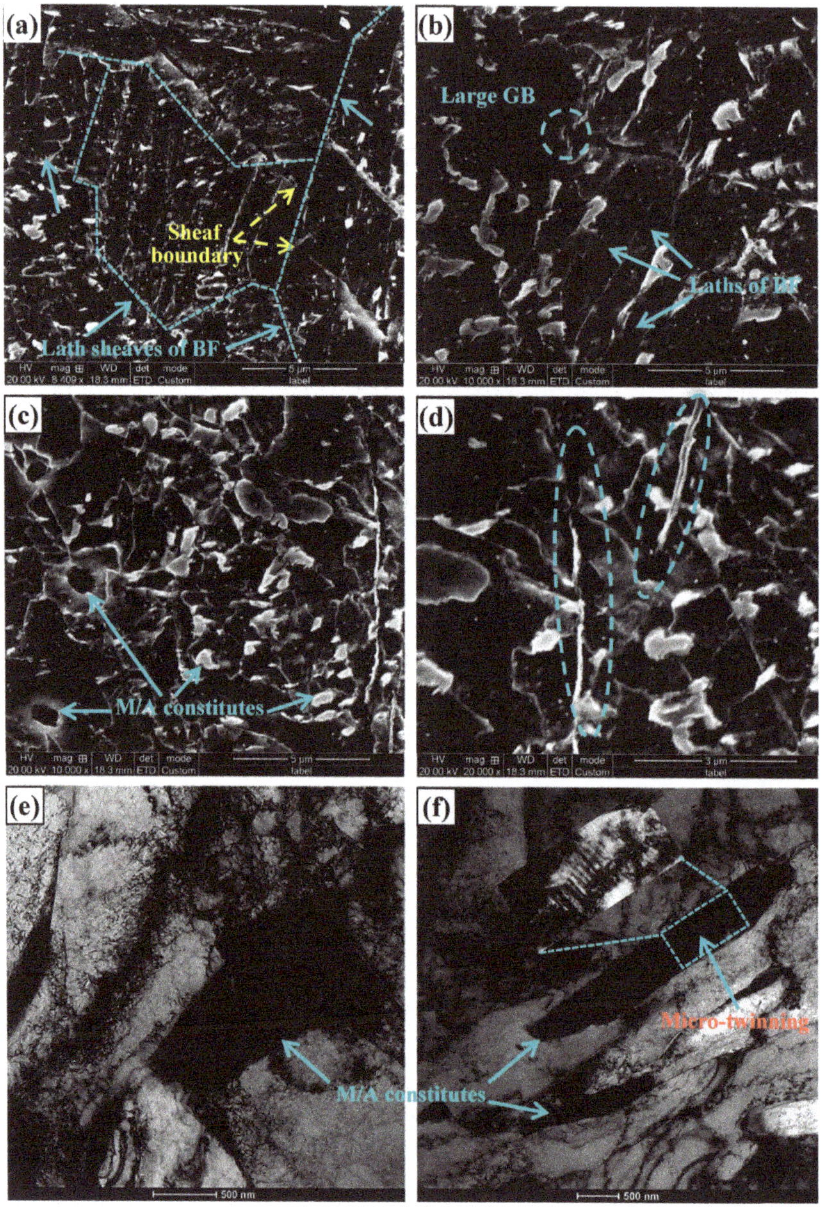

Figure 25. (**a–d**) High-magnification picture of secondary cracks in fracture cross-section of CVN specimen at −196 °C conditions; (**e**,**f**) TEM image of typical M/A component in granular bainite.

As a hard and brittle phase in the microstructure, the M/A constituent would also fall off (marked by the sky-blue arrow in Figure 25c) or crack (marked by the dashed ellipse in Figure 25d) during the deformation process, thus becoming the source of cracks. In particular, the strip M/A constituent is prone to micro-cracks and propagates cracks along itself. Figure 25e is a TEM image of a granular M/A constituent in granular bainite, which clearly shows the high density of dislocations inside and the entanglement of the dislocations caused by the surrounding stress field. The TEM image of the strip M/A constituent is then shown in Figure 25f, where micro-twins could be observed, which were twin martensitic transitions from the highly carbon-rich austenite, which had a higher hardness than the granular M/A in Figure 25e and was therefore more prone to micro-cracks.

4. Conclusions

In summary, the following conclusions have been drawn from the above investigation.

(1) Compared to CVN specimens, DWTT specimens of ultra-high-strength pipe steel had a variety of unconventional morphological features. The ductile fracture zone contained a large number of ridge-like protrusions and special-shaped dimples, where the ridge morphology was formed by a mass of dimples stretched along the direction of force. The brittle fracture zone contained a number of large-size fracture platforms. These features had the effect of consuming or hindering the propagation of cracks. Therefore, the DWTT fracture specimen provided more information about the material fracture process and could better reflect the actual process of the material fracturing.

(2) The sufficient flattening of the austenite in the non-recrystallized zone during the controlled rolling process was the basis for obtaining the excellent toughness of ultra-high-strength pipeline steel. During crack propagation, large angular deflections or frequent small transitions occurred when fine-grained granular bainite was encountered, which consumed more energy and improved toughness. When crack propagation encountered bainite ferrite, it would pass directly through the bainite lath composed of low-angle grain boundaries, but would deflect or stop at the lath sheaf boundaries.

(3) Large-sized granular bainite in ultra-high-strength pipe steel consisted of low-angle grain boundaries that did not effectively prevent crack propagation and were not conducive to improved toughness. Moreover, the M/A constituent in granular bainite aggregated, cracked or fell off, which was prone to lead to the formation of cracks and was also detrimental to toughness.

(4) In order to improve the toughness of ultra-high-strength pipe steel, the amount of large-sized granular bainite should be reduced while ensuring sufficient flattening of the deformed austenite.

Author Contributions: Conceptualization, J.G., B.L., X.Z., S.J., X.C., S.F. and D.Z.; Methodology, D.Z., B.L., X.Z., S.J., X.C., S.F., H.Z. and J.G.; Software and Experimental Validation, X.Z., B.L., S.J. and H.Z.; Formal Analysis and Investigation, S.J., B.L., X.Z. and X.C.; Resources, H.Z., S.F. and J.G.; Data Curation, S.F. and H.Z.; Writing-Original Draft Preparation, B.L.; Writing-Review and Editing, X.C.; B.L. and D.Z.; Supervision, S.F. and D.Z.; Project Administration, D.Z. and J.G.; Funding Acquisition, J.G. and D.Z.; All authors have read and agreed to the published version of the manuscript.

Funding: This research received no external funding.

Data Availability Statement: The original contributions presented in the study are included in the article, further inquiries can be directed to the corresponding author.

Conflicts of Interest: Authors Ba Li, Shujun Jia, Xiaoping Chen were employed by the company Central Iron and Steel Research Institute Company Limited, authors Ba Li, Song Fu, Dongliang Zhao, Haonan Zhang were employed by the company Zhejiang Innuovo Magnetics Co., Ltd., author Jie Guo was employed by the company PetroChina Planning and Engineering Institute. All authors declare that the research was conducted in the absence of any commercial or financial relationships that could be construed as a potential conflict of interest.

References

1. Koo, J.Y.; Luton, M.J.; Bangaru, N.V.; Petkovic, R.A.; Fairchild, D.P.; Petersen, C.W.; Asahi, H.; Hara, T.; Terada, Y.; Sugiyama, M.; et al. Metallurgical design of ultra-high strength steels for gas pipelines. *Int. J. Offshore Polar Eng.* **2004**, *14*, 2–10.
2. Witek, M. Possibilities of using X80, X100, X120 high-strength steels for onshore gas transmission pipelines. *J. Nat. Gas Sci. E* **2015**, *27*, 374–384. [CrossRef]
3. Duan, L.; Chen, Y.; Liu, Q.; Shu, J.; Cheng, C. Microstructures and mechanical properties of X100 pipeline steel strip. *J. Iron St Res. Int.* **2014**, *21*, 227–232. [CrossRef]
4. Yoo, J.Y.; Ahn, S.S.; Seo, D.H.; Song, W.H.; Kang, K.B. New development of high grade X80 to X120 pipeline steels. *Mater. Man Process.* **2011**, *26*, 154–160. [CrossRef]
5. Guillal, A.; Abdelbaki, N. The optimum use of high-strength steel in the construction of gas transmission pipeline. *Int. J. Stru* **2023**, *14*, 681–699. [CrossRef]
6. Ferino, J.; Fonzo, A.; Di Biagio, M.; Demofonti, G.; Spinelli, C.M.; Karamanos, S.A. Onshore pipeline high-grade steel for challen utilization. *Int. J. Offshore Polar Eng.* **2015**, *25*, 272–280. [CrossRef]
7. Lan, L.; Chang, Z.; Kong, X.; Qiu, C.; Zhao, D. Phase transformation, microstructure, and mechanical properties of X100 pipeli steels based on TMCP and HTP concepts. *J. Mater. Sci. Technol.* **2017**, *52*, 1661–1678. [CrossRef]
8. Li, X.C.; Zhao, J.X.; Wang, J.L.; Wang, X.L.; Liu, S.L.; Shang, C.J. Effect of Boundaries on Toughness in High-strength Low-all Steels from the View of Crystallographic Misorientation. *Mater. Lett.* **2019**, *259*, 126841. [CrossRef]
9. Liu, Q.Y.; Ba, L.; Jia, S.J.; Ren, Y. Toughening Mechanism and Controlling Technology for High-grade Low Temperature Pipeli Steels. *Angang Technol.* **2020**, *10*, 8–17.
10. Lei, Y.; Yang, W.; Siyasiya, C.W.; Tang, Z. Effect of Coiling Temperature on Microstructures and Precipitates in High-Streng Low-Alloy Pipeline Steel after Heavy Reduction during a Six-Pass Rolling Thermo-Mechanical Controlled Process. *Metals* **20** *14*, 249. [CrossRef]
11. Jiang, J.; Zhang, Z.; Guo, K.; Guan, Y.; Yuan, L.; Wang, Q. Effect of Nb Content on the Microstructure and Impact Toughness High-Strength Pipeline Steel. *Metals* **2023**, *14*, 42. [CrossRef]
12. Liu, D.; Dong, Y.; Li, R.; Jiang, J.; Li, X.; Wang, Z.; Zuo, X. Evaluation of Mechanical Properties and Microstructure of X70 Pipeli Steel with Strain-Based Design. *Metals* **2022**, *12*, 1616. [CrossRef]
13. Jiang, J.; Zhang, Z.; Zhao, L.; Guan, Y.; Yuan, L.; Wang, Q. Insight into the Role of Mo Content on the Microstructure and Imp Toughness of X80 Thick-Walled Low-Temperature Pipeline Steel. *Metals* **2023**, *13*, 1530. [CrossRef]
14. Li, X.C.; Zhao, J.X.; Jia, S.J.; Lu, G.Y.; Misra, R.D.K.; Liu, Q.Y.; Li, B. Ultrafine microstructure design of high strength pipeline st for low temperature service: The significant impact on toughness. *Mater. Lett.* **2021**, *303*, 130429. [CrossRef]
15. Qian, C.; Liang, J.; Huang, Y.; Kang, J.; Gianetto, J.J. A comparative study on constraint parameters for characterizing fractu toughness of X60 and X100 pipe steels. *Eng. Fail. Anal.* **2022**, *141*, 106617. [CrossRef]
16. Cheng, S.; Lv, R.; Cheng, F.; Han, B.; Fu, A.; Zhang, X.; Zhang, J.; Feng, Y.; Gao, H. Microstructure and properties of X1 large-deformability pipeline steel based on heating on-line partitioning. *Mater. Today Commun.* **2023**, *35*, 105771. [CrossRef]
17. Qi, X.; Huan, P.; Wang, X.; Di, H.; Shen, X.; Sun, Q.; Liu, Z.; He, J. Study on the mechanism of heat input on the grain bounda distribution and impact toughness in CGHAZ of X100 pipeline steel from the aspect of variant. *Mater. Charact.* **2021**, *179*, 1113 [CrossRef]
18. Singh, M.P.; Shukla, D.K.; Kumar, R.; Arora, K.S. The structural integrity of high-strength welded pipeline steels: A review. *Int Struct. Integr.* **2021**, *12*, 470–496. [CrossRef]
19. Bhardwaj, U.; Teixeira, A.P.; Soares, C.G. Burst strength assessment of X100 to X120 ultra-high strength corroded pipes. *Oce Eng.* **2021**, *241*, 110004. [CrossRef]
20. Hwang, B.; Shin, S.Y.; Lee, S.; Kim, N.; Kim, S.; Kang, K.B. Correlation of crack-tip opening angle for stable crack propagati with Charpy and drop-weight tear test properties in high-toughness API X70 pipeline steels. *Metall. Mater. Trans. A* **2006**, 371–380. [CrossRef]
21. Demofonti, G.; Mannucci, G.; Hillenbrand, H.G.; Harris, D. Evaluation of the suitability of X100 steel pipes for high pressure g transportation pipelines by full scale tests. *Int. Pipeline Conf.* **2004**, *41766*, 1685–1692.
22. Zhu, D.X.K.; Leis, B.N. Ductile-fracture arrest methods for gas-transmission pipelines using Charpy impact energy or DW energy. *J. Pipeline Sci. Eng.* **2013**, *12*, 259–272.
23. Wang, H.; Bao, Y.; Liu, Y.; Wang, M. Effect of microstructure and grain size on DWTT properties of 22 mm thick X80M hot roll steel strip. *Metall. Res. Technol.* **2019**, *116*, 220. [CrossRef]
24. Nastich, S.Y. Effect of bainite component morphology on the microstructure of X70 low-alloyed steel on thick plate cold resistan *Metallurgist* **2012**, *56*, 196–204. [CrossRef]
25. Shin, S.; Hwang, B.; Lee, S.; Kang, K.B. Effects of notch shape and specimen thickness on drop-weight tear test properties of A X70 and X80 line-pipe steels. *Metall. Mater. Trans. A* **2007**, *38*, 537–551. [CrossRef]
26. Hwang, B.; Kim, Y.M.; Lee, S.; Kim, N.; Yoo, J.Y. Correlation of rolling condition, microstructure, and low-temperature toughne of X70 pipeline steels. *Metall. Mater. Trans. A* **2005**, *36*, 1793–1805. [CrossRef]
27. Wang, H.; Bao, Y.; Wang, M.; Lin, L.; Li, Z.; Wei, X.D. Effect of inclusions on DWTT properties of thick gauge X80 hot rolled ste strip. *Chin. J. Eng. Sci.* **2018**, *40*, 1–10.

Zhao, J.; Hu, W.; Wang, X.; Kang, J.; Yuan, G.; Di, H.; Misra, R.D.K. Effect of microstructure on the crack propagation behavior of microalloyed 560 MPa (X80) strip during ultra-fast cooling. *Mater. Sci. Eng. A-Struct.* **2016**, *666*, 214–224. [CrossRef]
Gesing, M.A.; Simha, C.H.M.; Xu, S.; Tyson, W.R. Geometric and material property dependencies of the plastic rotation factor in the drop-weight-tear test. *Eng. Fract. Mech.* **2016**, *153*, 399–406. [CrossRef]
Sung, H.K.; Sohn, S.S.; Shin, S.Y.; Lee, S.; Kim, N.; Hwan, S.C. Effects of finish rolling temperature on inverse fracture occurring during drop weight tear test of API X80 pipeline steels. *Mater. Sci. Eng. A-Struct.* **2012**, *541*, 181–189. [CrossRef]
Hwang, B.; Lee, S.; Kim, Y.M.; Kim, N.; Yoo, J.Y.; Woo, C.S. Analysis of abnormal fracture occurring during drop-weight tear test of high-toughness line-pipe steel. *Mater. Sci. Eng. A-Struct.* **2004**, *368*, 18–27. [CrossRef]
Fujishiro, T.; Hara, T.; Aihara, S. Effect of plastic deformation on occurrence of abnormal fracture during DWTT. *Int. Pipeline Conf.* **2012**, *45141*, 159–165.
Yang, Z. The fracture during drop-weight tear test of high-performance pipeline steel and its abnormal fracture appearance. *Procedia Mater. Sci.* **2014**, *3*, 1591–1598. [CrossRef]
Kang, M.; Kim, H.; Lee, S.; Shin, S. Effects of dynamic strain hardening exponent on abnormal cleavage fracture occurring during drop weight tear test of API X70 and X80 linepipe steels. *Metall. Mater. Trans. A.* **2014**, *45*, 682–697. [CrossRef]
Majidi-Jirandehi, A.A.; Hashemi, S.H.; Ebrahiminejad, S.; Kheybari, M. Impact of crack propagation path and inclusion elements on fracture toughness and micro-surface characteristics of welded pipes in DWTT. *Mater. Res. Express* **2021**, *8*, 106504. [CrossRef]
ASTM E23-18; Standard Test Methods for Notched Bar Impact Testing of Metallic Materials. ASTM International: West Conshohocken, PA, USA, 2018.
GB T 8363-2018; Steel-Drop-Weight Tear Tests Method. China Standards Press: Beijing, China, 2018.
Sun, J.; Yang, C.; Guo, S.; Sun, X.; Ma, M.; Zhao, S.; Liu, Y. A novel process to obtain lamella structured low-carbon steel with bimodal grain size distribution for potentially improving mechanical property. *Mater. Sci. Eng. A* **2020**, *785*, 139339. [CrossRef]
Ye, X.; Cui, S.; Liu, T.; Ma, Q.; Liu, G.; Huang, Z.; Guo, J.; Yin, S. Microstructure Characterization and Strengthening Mechanism Analysis of X100 Pipeline Steel. *Coatings* **2023**, *13*, 706. [CrossRef]
Cho, L.; Bradley, P.E.; Lauria, D.S.; Martin, M.L.; Connolly, M.; Benzing, J.; Seo, E.J.; Findley, K.; Speer, J.G.; Slifka, A.J. Characteristics and mechanisms of hydrogen-induced quasi-cleavage fracture of lath martensitic steel. *Acta Mater.* **2021**, *206*, 116635. [CrossRef]
Niu, Y.; Jia, S.; Liu, Q.; Tong, S.; Li, B.; Ren, Y.; Wang, B. Influence of Effective Grain Size on Low Temperature Toughness of High-Strength Pipeline Steel. *Materials* **2019**, *12*, 3672. [CrossRef]
Venkatsurya, P.K.C.; Misra, R.D.K.; Manohar, M.; Hartmann, J.E. Toward interplay between substructure evolution, dislocation configuration, and yield strength in a microalloyed steel. *Mater. Sci. Eng. A* **2014**, *596*, 121–133. [CrossRef]
Lan, L.Y.; Qiu, C.L.; Song, H.Y.; Zhao, D.W. Correlation of martensite-austenite constituent and cleavage crack initiation in welding heat affected zone of low carbon bainitic steel. *Mater. Lett.* **2014**, *15*, 86–88. [CrossRef]
Huda, N.; Midawi, A.R.; Gianetto, J.; Lazor, R.; Gerlich, A.P. Influence of martensite-austenite (MA) on impact toughness of X80 line pipe steels. *Mater. Sci. Eng. A-Struct.* **2016**, *662*, 481–491. [CrossRef]

sclaimer/Publisher's Note: The statements, opinions and data contained in all publications are solely those of the individual thor(s) and contributor(s) and not of MDPI and/or the editor(s). MDPI and/or the editor(s) disclaim responsibility for any injury to ople or property resulting from any ideas, methods, instructions or products referred to in the content.

metals

Communication

Unraveling the Mechanism of Toughness Fluctuation in Ultra-High-Strength Casing from the Perspective of Crystallography

Zhan Gao [1,2], Xiaoming Dong [2], Jieru Yu [1], Junjie Tang [1], Yishuang Yu [3,*], Liangliang Wei [4,*], Shilong Liu [1,*] and Xuejun Jin [1,3]

1 Institute of Advanced Steels and Materials, School of Materials Science and Engineering, Shanghai Jiao Tong University, Shanghai 200240, China
2 Baosteel Research Institute, Baoshan Iron & Steel Co., Ltd., Shanghai 201900, China
3 School of Biomedical Engineering, Shanghai Jiao Tong University, Shanghai 200240, China
4 Spallation Neutron Source Science Center, Dongguan 523803, China
* Correspondence: yuyishuang@sjtu.edu.cn (Y.Y.); weill@ihep.ac.cn (L.W.); sliu6@sjtu.edu.cn (S.L.)

Abstract: The microstructure and impact toughness in an ultra-high-strength casing were investigated, attempting to reveal the reason for toughness fluctuations along the casing in terms of crystallography. The morphological structures at the head of the casing are lath bainite, while those at the end are granular bainite and lath bainite. The head exhibits a higher density of high-angle grain boundaries dominated by block boundaries and a higher impact toughness than the end, showing an inhomogeneous microstructure and causing toughness fluctuations in the casing. The higher density of block boundaries in the head resulted from weaker variant selection because of its higher cooling rate than the end. Therefore, the underlying reason for the toughness fluctuations lies in the varying densities of the block boundaries along the casing triggered by uneven cooling.

Keywords: casing; impact toughness; grain boundaries; variant pairing; phase transformation

1. Introduction

Casing, as a key structural component, is widely used in oil and gas exploration [1]. As easily exploitable shallow oil and gas resources are gradually drying up, explorations extend deeper [2]. The harsh service environments in deep oil and gas wells set high requirements for the strength, toughness and performance stability of the casing [3]. Generally, the impact energy of the casing should meet the requirements specified in the API Specification 5CT [1]. However, in some cases the microstructures show no notable differences, but the values of toughness exhibit sharp fluctuations along the casing, which seriously affects the product quality. Toughness fluctuation has increasingly become a bottleneck problem in the production of ultra-high-strength casings [2], and revealing the causes of toughness fluctuation has become a primary problem to be solved.

Toughness fluctuation is mainly related to two factors. One factor is casting defects, such as centerline segregation and casting inclusions [4,5]. Centerline segregation usually causes the formation of band microstructures in the heavy plate, which alters the carbide precipitation behavior and lowers the toughness [6]. Additionally, the casting inclusion, like bifilms and oxide inclusions, may also affect the toughness [7]. The other factor is the microstructural heterogeneity, which is usually induced by an uneven rolling deformation and cooling rate [8–10]. Microstructural heterogeneity usually exists across the thickness direction of heavy plates or along the axis of the steel pipe [11,12]. With the improvement in the casting level, inclusions have been well controlled and are no longer the main cause of toughness fluctuations. For low-alloy casing steel, due to its low Mn content (~1.0 wt.%), the degree of Mn segregation is relatively light. Therefore, the casting defects can be ignored for casing steel, and microstructural heterogeneity becomes the main reason for toughness fluctuations. An in-depth analysis of the microstructural heterogeneity of casing steel helps to reveal the mechanism of toughness fluctuations.

The microstructures of ultra-high-strength casings are typical lath structures, i.e., martensite/bainite. Lath structures have a hierarchical microstructure which can be categorized into multiscale crystallographic units like prior austenite grains, packets, blocks, sub-blocks and laths [13,14]. A prior austenite grain is composed of several martensite packets with the same habit plane; each packet is then divided into a martensite block, which contains several laths with the same crystallographic orientation. Based on the crystallographic orientation relationship between austenite and bainite, there exist 24 variants transformed from a single austenite grain in low carbon steels, which can be divided into four closed-packed (CP) groups or three Bain groups [15]. Each CP group is made up of six variants sharing the same parallel relationship of closed-packed planes with austenite (CP1: $(111)\gamma//(011)\alpha$, CP2: $(1-11)\gamma//(011)\alpha$, CP3: $(-111)\gamma//(011)\alpha$ and CP4: $(11-1)\gamma//(011)\alpha$). The Bain group consists of eight variants with the same Bain correspondence (B1: $[001]\gamma//[001]\alpha$, B2: $[100]\gamma//[110]\alpha$ and B3: $[010]\gamma//[-110]\alpha$). The boundaries that separate these microstructural units are different in nature and have different misorientation angles across them. In general, prior austenite grain (PAG), packet and block boundaries are high-angle grain boundaries, and lath boundaries are low-angle grain boundaries in martensitic/bainitic structures [13,14,16]. Earlier studies show that refining the PAGs, packets and blocks has a beneficial effect on the enhancement of impact toughness because of the effectiveness in hindering the cleavage crack propagation [17–20]. Studies on the reasons for toughness fluctuations depending on the microstructural heterogeneity from a crystallographic perspective are relatively scarce. Therefore, a crystallographic analysis of the microstructure of the casing can greatly help to elucidate the relationship between crystallographic features and impact toughness, and thus reveal the causes of toughness variation.

In this report, the microstructural heterogeneity in a high-strength casing in terms of crystallography was studied to uncover the mechanism of toughness fluctuation.

2. Materials and Methods

The material investigated in this work is an industrial-produced 155 ksi grade casing with a chemical composition of Fe-0.16C-0.25Si-1.00Mn-2.0(Cr + Mo)-0.09(Nb + V) (wt.%). After hot rolling, the casing was firstly reaustenitized at 900 °C for 2 h, subsequently water-quenched to form lath structures and then tempered at 600 °C for 2 h. These processes were carried out on a continuous industrial production line.

Specimens for the tensile and impact tests and microstructure observation were cut from the head and end of the casing along the rolling direction. The tensile properties were measured at room temperature using flat dog-bone-shaped tensile specimens with a thickness of 12 mm and a gauge section of 50×25 mm^2 according to the GB/T 228.1−2021 specification [21]. The tensile tests were conducted using a universal testing machine (WDW-200, Changchun, China) at a strain rate of 1.0×10^{-3} s^{-1}. Impact toughness was measured at 0 °C using Charpy V-notched impact specimens according to the GB/T 229−2020 standard [22], and the impact tests were carried out on a pendulum impact testing machine (PTM2200−D1, Shenzhen, China). Three specimens were tested for each condition.

The samples for microstructure observation were mounted and mechanically polished. After being etched using 4% nital solution and electropolished using a solution of 10% perchloric acid in ethanol, the samples were characterized by using a TESCAN MIRA3 field emission scanning electron microscope (TESCAN, Brno, Czech Republic) equipped with an Oxford electron backscatter diffraction (EBSD) detector (Oxford Instruments, Oxford, UK). EBSD data were obtained under the following conditions: accelerating voltage, 20 kV; working distance, 15 mm; tilt angle, 70°; step size, 100 nm. HKL Channel 5 software (Oxford Instruments, Oxford, UK) and MATLAB® R2017b software (MathWorks, Natick, MA, USA) were utilized to visualize and digitize the crystallographic features. To reveal the features of lath structures and carbides, thin foils with a diameter of 3 mm were prepared, and TEM analysis was conducted using a Talos F20 transmission electron microscope (Thermo Fisher, Waltham, MA, USA) operated at an accelerating voltage of 200 kV.

3. Results and Discussion

Figure 1 displays the tensile and impact properties of the head and end samples. The samples exhibit similar tensile properties with a yield strength (YS) of ~1100 MPa, an ultimate tensile strength (UTS) of ~1200 MPa and a total elongation (TEL) of ~21%. However, the end shows a lower 0 °C impact toughness with a reduction of ~41% compared with that of the head, implying that the impact toughness varies along the casing. Obviously, the impact property cannot meet the toughness requirement of a 155 ksi grade casing.

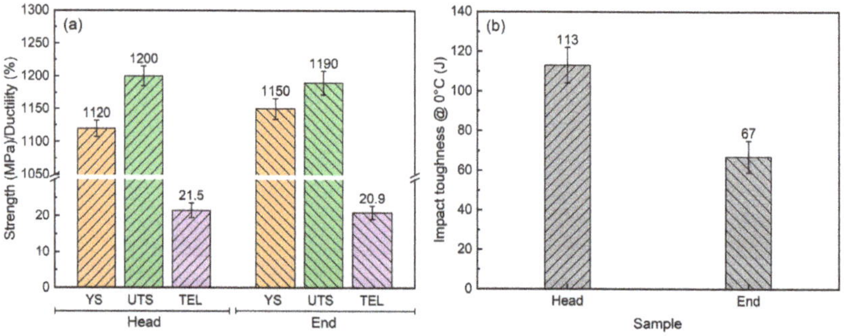

Figure 1. Tensile properties (**a**) and impact toughness (**b**) of the head and end samples. YS, UTS and TEL represent yield strength, ultimate tensile strength and total elongation, respectively.

The SEM micrographs and TEM images of the head and end samples are depicted in Figure 2. The morphological structures at the head are characterized by typical lath bainite (LB), while those at the end are composed of LB with some granular bainite (GB) (Figure 2a,d). In addition, the lath boundary and prior austenite grain boundary (PAGB) are decorated with bright and fine carbides in both samples. In the TEM images (Figure 2b,c,e,f), the width of structures at the head is ~100–200 nm, which is finer compared with that at the end with a width thicker than 200 nm. Carbides with a size of ~50–100 nm can be observed in both samples.

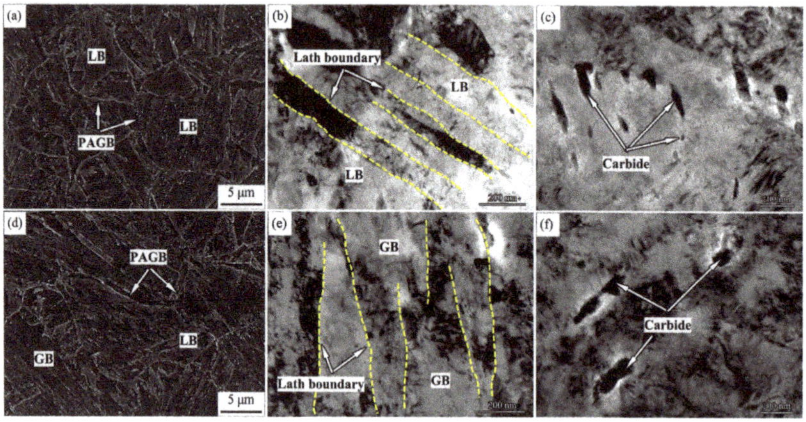

Figure 2. SEM micrographs (**a**,**d**) and TEM images (**b**,**c**,**e**,**f**) depicting the microstructure and carbide distribution in the head (**a**–**c**) and end (**d**–**f**) samples.

Figure 3 displays the band contrast (BC) and inverse pole figure (IPF) maps of the head and end samples. Notable differences in the crystallographic features are observed in the samples. In the BC maps, the boundaries are divided into three types according to the misorientation (θ), i.e., low-angle grain boundaries (LAGBs, $5^\circ < \theta < 15^\circ$), high-angle grain boundaries (HAGBs, $15^\circ < \theta < 45^\circ$) and high-angle grain boundaries (HAGBs, $\theta > 45^\circ$), and are highlighted by white, black and red lines, respectively. Compared with the end, the head exhibits straighter boundaries, most of which are equivalent boundaries of V1/V2 [23]. The head has finer microstructure and higher density of high angle grain boundaries (HAGBs). Previous studies show that toughness is influenced by the distribution of special misorientation angle (SMA) between the {1 0 0} planes and SMA between the {1 1 0} planes [20,24]. The boundaries are vital to the crack initiation and propagation, which are influenced by the boundaries with a high {1 1 0}-SMA and a high {1 0 0}-SMA, respectively. At the stage of crack initiation, boundaries with high {1 1 0}-SMAs will hinder the dislocation gliding, thus lowering the stress concentration at the notch. As a result, larger plastic deformation forms, and more impact energy is obtained. Once a brittle crack forms and begins to propagate, boundaries with high {1 0 0}-SMAs become the main influencing factor by hindering the crack propagation. Boundaries with high {1 0 0}-SMAs are usually HAGBs. Therefore, the boundaries, especially the HAGBs, contribute to the enhancement of impact toughness [20,24–28].

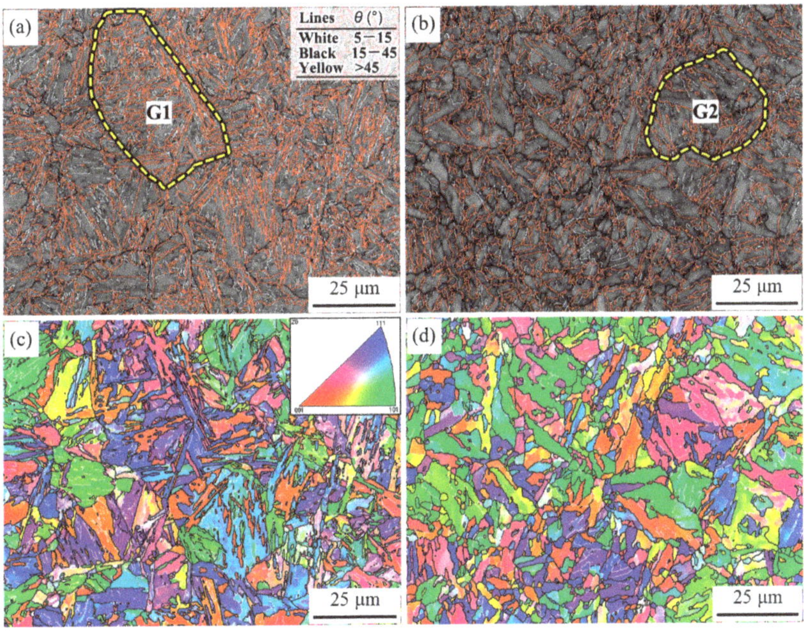

Figure 3. BC (**a**,**b**) and IPF (**c**,**d**) maps depicting the microstructure and boundary distribution in the head (**a**,**c**) and end (**b**,**d**) samples.

To further analyze the microstructural heterogeneity of the samples, an inspection was conducted to visualize and digitize the crystallographic features. Figure 4 shows the structures of representative grains highlighted with yellow dotted lines in Figure 3a,b. One can see from the GB and IPF maps that the G1 grain from the head exhibits a higher density of HAGBs and finer structures than the G2 grain from the end. In this study, the 24 variants transformed from the representative grain were classified as four CP groups and three Bain groups which were highlighted with four and three colors, respectively. It is noticed from the CP and Bain group maps that the boundaries between variants in the same Bain region are LAGBs (white lines), while the boundaries between variants belonging to different Bain groups are HAGBs (black lines). Interestingly, most of the CP regions in the G1 grain are occupied by different fine Bain groups, implying that variants from distinct Bain groups are preferentially formed. In contrast, the CP regions in the G2 grain are inclined to an arrangement with the same coarse Bain groups, which is detrimental to the formation of HAGBs.

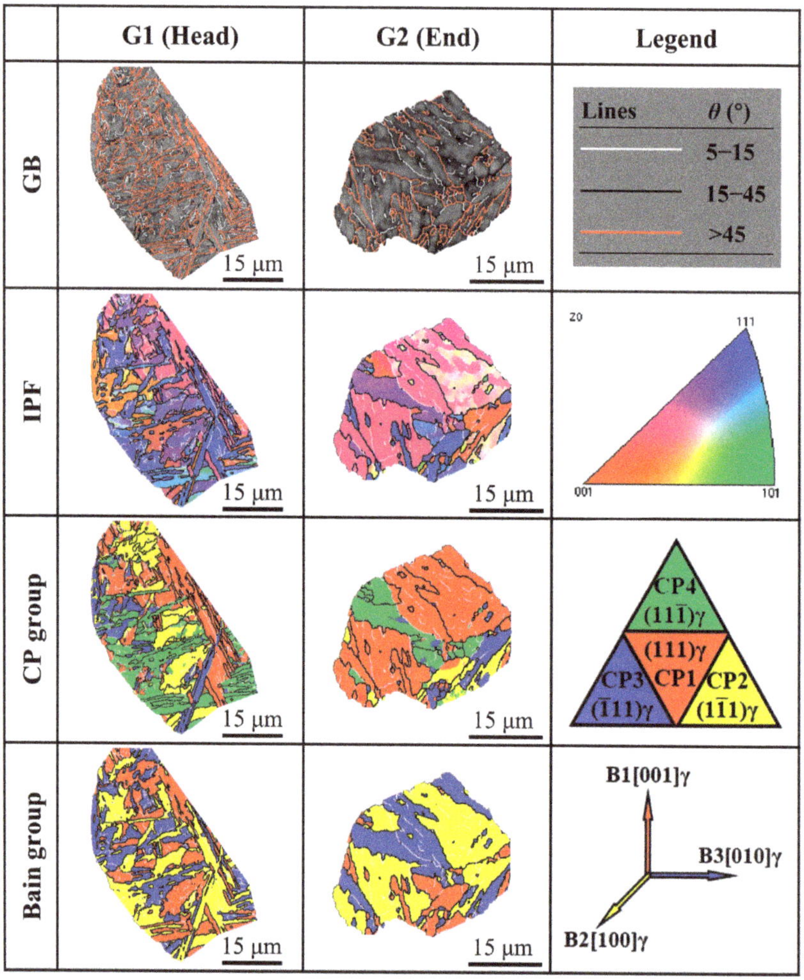

Figure 4. Visualization of the crystallographic features in the representative grains: GB maps, IPF and corresponding CP and Bain maps of the G1 and G2 grains.

The HAGBs and arrangements of CP and Bain groups within the grains are associated with variant selection [29–32]. Based on the actual orientation relationship, the fractions of the variant in the G1 and G2 grains were calculated quantitatively according to the method reported in refs. [33,34], and the results are pictured in Figure 5. Horizontal dash–dot lines stand for the fraction of the random distribution of variants (i.e., 1/24), which denotes that no variant selection occurs. Clearly, a total of 24 variants from V1 to V24 were all formed in both grains. The G1 grain from the head shows a very weak variant selection verified by the random distribution of 24 variants, while the G2 grain from the end formed predominant variants (i.e., V10, V16, V18 and V21), showing a strong variant selection. Previous studies suggest that a high cooling rate can weaken the variant selection and contribute to the formation of different variants and the refinement of packets and blocks [8,23,35]. In this work, the different variant pairing tendencies in the head and the end resulted from uneven cooling along the casing. It can be inferred that the cooling rate at the head should be higher than that at the end since the variant selection in the head is weaker.

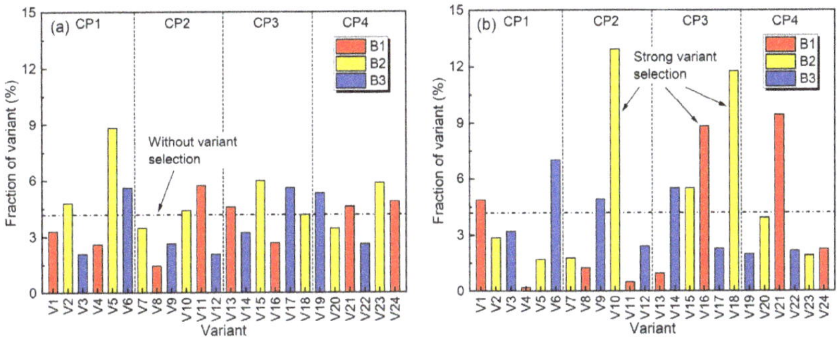

Figure 5. Digitization of the crystallographic features in the representative grains: fractions of the variant in the G1 (**a**) and G2 (**b**) grains.

For better comprehension of the relationship between intervariant boundaries and impact toughness, the intervariant boundaries were further categorized into block, sub-block and packet boundaries according to the boundary types [13,15], and the densities of the intervariant boundaries were calculated. As shown in Figure 6a, the head exhibits a significantly higher density of block boundaries from different Bain groups than the end, which is a result of its weaker variant selection [8,36]. Previous studies have shown that a lower starting temperature of transformation can be triggered by a higher cooling rate, and thus a greater driving force for phase transformation can be obtained [8,35,37]. In this work, the higher cooling rate in the head provided a greater driving force, more variants from distinct Bain groups formed and the block width became finer, thus accommodating greater transformation strain [15,38,39]. Figure 6b summarizes the boundary density and impact toughness. The density of block boundaries shows a similar tendency with that of the impact toughness, suggesting that a HAGB dominated by block boundaries is the key factor affecting the toughness, consistent with the results in refs. [9,27,40–42]. Therefore, the high cooling rate in the head weakened the variant selection, resulting in the high density of block boundaries and impact toughness.

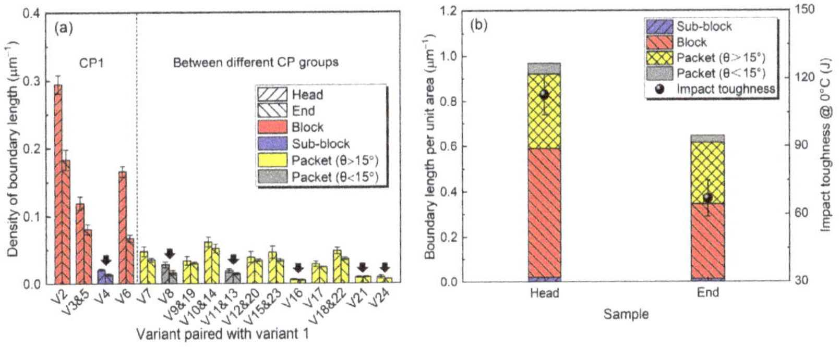

Figure 6. Density of intervariant boundaries (**a**) and relationship between boundary density and impact toughness (**b**) of the samples. Black arrows point out the intervariant relations belonging to the same Bain group.

4. Conclusions

The mechanism responsible for the toughness fluctuation in ultra-high-strength casings was elucidated from the aspect of crystallography. The head of the casing exhibits a higher impact toughness and density of HAGBs dominated by block boundaries and weaker variant selection compared with the end. The variation in the impact toughness between the head and the end is related to the different cooling rates along the casing. The high cooling rate at the head can provide a large driving force and weaken the variant selection, which results in the high density of block boundaries and impact toughness. Thus, the varying density of block boundaries provoked by uneven cooling is responsible for toughness fluctuations in the casing.

Author Contributions: Conceptualization, Y.Y., L.W. and S.L.; methodology, Y.Y., L.W. and S.L.; software, Y.Y.; validation, Z.G., Y.Y. and S.L.; formal analysis, Z.G., X.D., Y.Y. and L.W.; investigation, Z.G., X.D., J.Y., Y.Y. and L.W.; resources, Z.G.; data curation, Y.Y.; writing—original draft preparation, Z.G. and Y.Y.; writing—review and editing, J.Y., J.T., Y.Y. and L.W.; visualization, Z.G. and Y.Y.; supervision, X.J.; project administration, X.J.; funding acquisition, Z.G., X.D., Y.Y. and S.L. All authors have read and agreed to the published version of the manuscript.

Funding: This research was funded by the National Key Research and Development Program of China (No. 2023YFB3711700); the China Postdoctoral Science Foundation (No. 2023M732192); the National Natural Science Foundation of China (No. 52101045); and the National Postdoctoral Foundation of China (Nos. BX20180186 and 2019M661498).

Data Availability Statement: The data presented in this study are available on request from the corresponding author due to legal and ethical reasons.

Conflicts of Interest: Z.G. and X.D. were employed by the Baoshan Iron & Steel Co., Ltd. The remaining authors declare that the research was conducted in the absence of any commercial or financial relationships that could be construed as a potential conflict of interest.

References

1. Li, H.; Han, L.; Zhang, W. Demand for and development of hi-performance OCTG. *Steel Pipe* **2009**, *38*, 1–9. [CrossRef]
2. Feng, Y.; Li, H.; Han, L.; Zhang, Z.; Zhang, C.; Cheng, H.; Bai, Z.; Tian, Y.; Shi, X.; Xie, J.; et al. Progress and prospects manufacturing technology for oil country tubular goods in China. *Pet. Sci. Bull.* **2022**, *7*, 229–241. [CrossRef]
3. Chen, H.; Yu, Y.S.; Wu, B.B.; Shang, C.J. Metallurgical design and high strength-toughness mechanism of ultrahigh strengt toughness oil pipeline steel. *Kang T'ieh/Iron Steel* **2019**, *54*, 96–103. [CrossRef]
4. Costa e Silva, A.L.V.d. Non-metallic inclusions in steels—Origin and control. *J. Mater. Res. Technol.* **2018**, *7*, 283–299. [CrossRe
5. Liu, Z.P.; Xie, Z.J.; Luo, D.; Zhou, W.H.; Guo, H.; Shang, C.J. Influence of central segregation on the welding microstructure ar properties of FH40 cryogenic steel. *Gongcheng Kexue Xuebao/Chin. J. Eng.* **2023**, *45*, 1335–1341. [CrossRef]
6. Han, P.; Liu, Z.; Xie, Z.; Wang, H.; Jin, Y.; Wang, X.; Shang, C. Influence of band microstructure on carbide precipitation behavi and toughness of 1 GPa-grade ultra-heavy gauge low-alloy steel. *Int. J. Miner. Metall. Mater.* **2023**, *30*, 1329–1337. [CrossRef]
7. Xiong, Z.; Liu, S.; Wang, X.; Shang, C.; Misra, R.D.K. Relationship between crystallographic structure of the Ti_2O_3/MnS compl inclusion and microstructure in the heat-affected zone (HAZ) in steel processed by oxide metallurgy route and impact toughne *Mater. Charact.* **2015**, *106*, 232–239. [CrossRef]
8. Yu, H.; Yu, Y.; Wang, Z.; Li, F.; Hu, B.; Liu, S. On the variant pairing in transformation product of high strength low alloy ste depending on cooling rate. *Mater. Lett.* **2022**, *326*, 132953. [CrossRef]
9. Liu, Z.C.; Wang, X.L.; Yu, Y.S.; Shang, C.J. Study on toughening mechanism of high strength steel and its relationship wi substructure. *J. Iron Steel Res.* **2020**, *32*, 1093–1101. [CrossRef]
10. Bai, P.; Shang, C.; Wu, H.-H.; Ma, G.; Wang, S.; Wu, G.; Gao, J.; Chen, Y.; Zhang, J.; Zhu, J.; et al. A review on the advance low-temperature toughness in pipeline steels. *J. Mater. Res. Technol.* **2023**, *25*, 6949–6964. [CrossRef]
11. Huang, S.; Yu, Y.S.; Wang, Z.Q.; Su, S.; Chen, K.; Yuan, S.F.; Xie, Z.J.; Shang, C.J. Crystallographic insights into the role of nickel hardenability of wear-resistant steels. *Mater. Lett.* **2022**, *306*, 130961. [CrossRef]
12. Chen, Q.-Y.; Zhang, W.-N.; Wang, P.-J.; Mao, Q.-J.; Liu, Z.-Y. Crystallography of transformation products with different coolir rates in low-carbon alloy steel and its effect on low-temperature toughness uniformity of heavy plates. *J. Mater. Res. Technol.* **202** *28*, 2077–2085. [CrossRef]

Morito, S.; Tanaka, H.; Konishi, R.; Furuhara, T.; Maki, T. The morphology and crystallography of lath martensite in Fe-C alloys. *Acta Mater.* **2003**, *51*, 1789–1799. [CrossRef]
Morito, S.; Huang, X.; Furuhara, T.; Maki, T.; Hansen, N. The morphology and crystallography of lath martensite in alloy steels. *Acta Mater.* **2006**, *54*, 5323–5331. [CrossRef]
Takayama, N.; Miyamoto, G.; Furuhara, T. Effects of transformation temperature on variant pairing of bainitic ferrite in low carbon steel. *Acta Mater.* **2012**, *60*, 2387–2396. [CrossRef]
Luo, H.; Wang, X.; Liu, Z.; Yang, Z. Influence of refined hierarchical martensitic microstructures on yield strength and impact toughness of ultra-high strength stainless steel. *J. Mater. Sci. Technol.* **2020**, *51*, 130–136. [CrossRef]
Morris, J.J.W.; Lee, C.S.; Guo, Z. The Nature and Consequences of Coherent Transformations in Steel. *ISIJ Int.* **2003**, *43*, 410–419. [CrossRef]
Wang, C.; Wang, M.; Shi, J.; Hui, W.; Dong, H. Effect of microstructural refinement on the toughness of low carbon martensitic steel. *Scr. Mater.* **2008**, *58*, 492–495. [CrossRef]
Chatterjee, A.; Chakrabarti, D.; Moitra, A.; Mitra, R.; Bhaduri, A.K. Effect of deformation temperature on the ductile–brittle transition behavior of a modified 9Cr–1Mo steel. *Mater. Sci. Eng. A* **2015**, *630*, 58–70. [CrossRef]
Li, X.C.; Zhao, J.X.; Cong, J.H.; Misra, R.D.K.; Wang, X.M.; Wang, X.L.; Shang, C.J. Machine learning guided automatic recognition of crystal boundaries in bainitic/martensitic alloy and relationship between boundary types and ductile-to-brittle transition behavior. *J. Mater. Sci. Technol.* **2021**, *84*, 49–58. [CrossRef]
GB/T 228.1-2021; Metallic Materials—Tensile Testing—Part I: Method of Test at Room Temperature. China Standard Press: Beijing, China, 2021.
GB/T 229−2020; Metallic Materials—Charpy Pendulum Impact Test Method. China Standard Press: Beijing, China, 2020.
Huang, S.; Wu, B.B.; Wang, Z.Q.; Yu, Y.S.; Wang, C.S.; Yan, L.; Li, X.C.; Shang, C.J.; Misra, R.D.K. EBSD study on the significance of carbon content on hardenability. *Mater. Lett.* **2019**, *254*, 412–414. [CrossRef]
Li, X.; Zhao, J.; Wang, J.; Wang, X.; Liu, S.; Shang, C. Effect of boundaries on toughness in high-strength low-alloy steels from the view of crystallographic misorientation. *Mater. Lett.* **2020**, *259*, 126841. [CrossRef]
Gourgues, A.F. Electron backscatter diffraction and cracking. *Mater. Sci. Technol.* **2002**, *18*, 119–133. [CrossRef]
Li, X.; Zhao, J.; Dong, L.; Devesh Kumar Misra, R.; Wang, X.; Wang, X.; Shang, C. The significance of coherent transformation on grain refinement and consequent enhancement in toughness. *Materials* **2020**, *13*, 5095. [CrossRef]
Yu, Y.-S.; Wang, Z.-Q.; Wu, B.-B.; Zhao, J.-X.; Wang, X.-L.; Guo, H.; Shang, C.-J. Tailoring variant pairing to enhance impact toughness in high-strength low-alloy steels via trace carbon addition. *Acta Metall. Sin.* **2021**, *34*, 755–764. [CrossRef]
Fan, E.; Li, Y.; You, Y.; Lü, X. Effect of crystallographic orientation on crack growth behaviour of HSLA steel. *Int. J. Miner. Metall. Mater.* **2022**, *29*, 1532–1542. [CrossRef]
Yu, Y.S.; Wang, Z.Q.; Wu, B.B.; Rong, X.Q.; Wei, L.J.; Yuan, S.F.; Guo, H.; Shang, C.J. New insight into the hardenability of high strength low alloy steel from the perspective of crystallography. *Mater. Lett.* **2021**, *292*, 129624. [CrossRef]
Ståhlkrantz, A.; Hedström, P.; Sarius, N.; Borgenstam, A. Effect of carbon content on variant pairing in bainitic low alloy steel. *Metall. Mater. Trans. A* **2022**, *53*, 3418–3427. [CrossRef]
Wang, B.; Chen, N.; Cai, Y.; Guo, W.; Wang, M. Effect of crystallographic features on low-temperature fatigue ductile-to-brittle transition for simulated coarse-grained heat-affected zone of bainite steel weld. *Int. J. Fatigue* **2023**, *170*, 107523. [CrossRef]
Yin, T.W.; Shen, Y.F.; Jia, N.; Li, Y.J.; Xue, W.Y. Controllable selection of martensitic variant enables concurrent enhancement of strength and ductility in a low-carbon steel. *Int. J. Plast.* **2023**, *168*, 103704. [CrossRef]
Miyamoto, G.; Takayama, N.; Furuhara, T. Accurate measurement of the orientation relationship of lath martensite and bainite by electron backscatter diffraction analysis. *Scr. Mater.* **2009**, *60*, 1113–1116. [CrossRef]
Miyamoto, G.; Iwata, N.; Takayama, N.; Furuhara, T. Mapping the parent austenite orientation reconstructed from the orientation of martensite by EBSD and its application to ausformed martensite. *Acta Mater.* **2010**, *58*, 6393–6403. [CrossRef]
Wu, B.B.; Wang, Z.Q.; Yu, Y.S.; Wang, X.L.; Shang, C.J.; Misra, R.D.K. Thermodynamic basis of twin-related variant pair in high strength low alloy steel. *Scr. Mater.* **2019**, *170*, 43–47. [CrossRef]
Liu, Z.; Yu, Y.; Yang, J.; Wang, Z.; Guo, H.; Shang, C. Morphology and crystallography analyses of HSLA steels with hardenability enhanced by tailored C–Ni collocation. *Metals* **2022**, *12*, 32. [CrossRef]
Long, X.; Liu, W.; Zhu, R.; Zhang, Y.; Zhang, F.; Yang, Z.; Li, Y. Effect of the cooling rate in the medium temperature zone on the phase transformation and microstructure of carbide-free bainitic steel. *J. Mater. Res. Technol.* **2024**, *29*, 50–66. [CrossRef]
Kawata, H.; Sakamoto, K.; Moritani, T.; Morito, S.; Furuhara, T.; Maki, T. Crystallography of ausformed upper bainite structure in Fe–9Ni–C alloys. *Mater. Sci. Eng. A* **2006**, *438–440*, 140–144. [CrossRef]
Xi, X.; Wu, T.; Wang, S.; Wang, J.; Wang, G.; Chen, L. Research on the mechanism of Cu addition on the grain boundary characteristics and toughness in CGHAZ of low carbon low alloy steel. *J. Mater. Res. Technol.* **2023**, *23*, 577–586. [CrossRef]
Li, X.; Lu, G.; Wang, Q.; Zhao, J.; Xie, Z.; Misra, R.D.K.; Shang, C. The effects of prior austenite grain refinement on strength and toughness of high-strength low-alloy steel. *Metals* **2022**, *12*, 28. [CrossRef]

41. Wang, X.L.; Xie, Z.J.; Wang, Z.Q.; Yu, Y.S.; Wu, L.Q.; Shang, C.J. Crystallographic study on microstructure and impact toughne of coarse grained heat affected zone of ultra-high strength steel. *Mater. Lett.* **2022**, *323*, 132552. [CrossRef]
42. Qi, X.; Huan, P.; Wang, X.; Di, H.; Shen, X.; Sun, Q.; Liu, Z.; He, J. Study on the mechanism of heat input on the grain bounda distribution and impact toughness in CGHAZ of X100 pipeline steel from the aspect of variant. *Mater. Charact.* **2021**, *179*, 1113 [CrossRef]

Disclaimer/Publisher's Note: The statements, opinions and data contained in all publications are solely those of the individu author(s) and contributor(s) and not of MDPI and/or the editor(s). MDPI and/or the editor(s) disclaim responsibility for any injury people or property resulting from any ideas, methods, instructions or products referred to in the content.

metals

MDPI

icle

ffect of a Gradient Temperature Rolling Process on the licrostructure and Mechanical Properties of the Center of ltra-Heavy Plates

ghua Cong [1], Jingxiao Zhao [1], Xuemin Wang [1,*] and Zhongwen Wu [2]

[1] Collaborative Innovation Center of Steel Technology, University of Science and Technology Beijing, Beijing 100083, China; congjinghua@126.com (J.C.); zhjingxiao@yeah.net (J.Z.)
[2] Xiangtan Iron & Steel Group Co., Ltd., Xiangtan 411200, China; m15973218185@163.com
* Correspondence: wxm@mater.ustb.edu.cn

Abstract: As there is a small amount of deformation in the center during the rolling process of ultra-heavy plates, it is extremely easy to cause poor mechanical properties in the center. Increasing the deformation in the center is the most feasible method to eliminate the deformation effects in the cross-section of ultra-heavy plates. In this study, the gradient temperature rolling (GTR) process is compared with the traditional uniform temperature rolling (UTR) process. It is found that the GTR process can significantly increase the deformation in the center and thereby refine the grains. The room temperature tensile test and instrumented Charpy impact test are used to test the strength at room temperature and impact energy at low temperature. Combined with the obtained impact load/energy displacement curve, the deformation and damage process under impact load are analyzed. The microstructure morphology and impact fracture obtained by different rolling processes in the center are analyzed by experimental methods such as OM, SEM, EBSD, etc. The prior austenite grain (PAG) boundary morphology is analyzed and the densities of grain boundaries are statistically quantified. The results showed that the strength, plasticity, and low-temperature toughness of the GTR process are improved compared to the UTR process, with increased dislocation density in the center microstructure, the density of PAG boundaries, and the density of packet boundaries. The size of the PAG in the center is refined by ~49%, the density of PAG boundaries increased by ~140%, the density of high-angle packet boundaries increased by ~39%, and the density of low-angle packet boundaries increased by ~49%. The crack propagation in the instrumented Charpy impact test of the GTR process showed stable expansion, indicating a ductile fracture compared to the semi-brittle fracture of the UTR process. The densities of PAG boundaries and high-angle packet boundaries are the most important factors affecting the strength and low-temperature toughness.

Keywords: gradient temperature rolling process; uniform temperature rolling process; ultra-heavy plate; prior austenite grain boundary; high-angle packet boundary

1. Introduction

The application prospects of an ultra-heavy plate in major technical equipment manufacturing fields such as marine engineering, pressure vessels, nuclear power, wind power, military, heavy machinery, mold manufacturing, etc., are enormous and the functional requirements for its welding performance, high strength, low-temperature toughness, etc., are also increasing [1–5]. At present, the primary problem affecting the quality of ultra-heavy plates is the uneven distribution of microstructures and mechanical properties in the thickness direction (cross-sectional effect) [4,6–9]. The difference in the properties of the center of ultra-heavy plates compared to other locations is mainly due to the differences in transformation and precipitates caused by temperature and strain fields, especially the microstructure, grain boundaries, and precipitates.

The elimination of the cross-sectional effect of ultra-heavy plates is mainly achieved through the TMCP (thermomechanical control) process and heat treatment process [4,5,8,10–12]. Due to the limitation of thermal conductivity, the center is prone to insufficient cooling capacity, resulting in uneven microstructural properties during the heat treatment process, especially the appearance of granular bainite, which affects the strength and toughness. The TMCP process is a key technology for the production of ultra-heavy plates with low cost and high efficiency. However, due to the uneven deformation during rolling and the uneven distribution of the temperature field during cooling, it is bound to cause problems such as uneven microstructural properties in the thickness direction, difficulty in quality controlling, and large differences in section characteristics, mainly due to poor toughness of the central microstructure, which affects the application of the material [8]. The GTR process is an effective method to solve this problem. It uses a laminar cooling/rapid cooling process to form a temperature gradient in the thickness direction of the slab before rolling, with the center temperature being far higher than the surface temperature [4,8]. The lower deformation resistance of the higher center temperature makes deformation easier to penetrate into the center, resulting in more uniform deformation and recrystallization of the steel plate in the thickness direction [8,13].

Research on the GTR process mainly focuses on the simulation of temperature gradients during the rolling process, the deformation penetration at different locations after rolling, and the impact on mechanical properties. However, there is a lack of research on microstructural refinement, especially the impact of various types of grain boundary densities on the strength and toughness. This study compares the GTR process with the traditional UTR process, distinguishes different grain boundary types, and analyzes the impact of different rolling processes on the microstructure and mechanical properties of the center from dislocation density, grain boundary types, and grain boundary density.

2. Materials and Methods

Table 1 shows the chemical composition of the experimental steel. The steel was smelted, forged, and cut into a rectangular samples with 120 mm thickness, 120 mm width, and 200 mm length. The samples were soaked at 1200 °C for 2 h to dissolve microalloying elements and subsequently rolled to 50 mm via multiple passes, as shown in Figure 1. Both rolling processes were performed in a continuous manner without changing the rolling direction. The two as-rolled steel samples are henceforth referred as UTR (uniform temperature rolling) and GTR (gradient temperature rolling), respectively. Research has shown that rapid cooling of the surface before rolling can create a temperature gradient between the center and the surface (with lower temperatures closer to the surface), with the center temperature still around 1100 °C [4,6–9]. The surface temperature decreases as the rapid cooling time increases. The center–surface temperature difference in UTR processes is very small. The start rolling temperature of the UTR process in this study is ~1100 °C on the surface and the center temperature is also close to 1100 °C [8]. The resistance to deformation of the surface and center are close. The GTR process involves water cooling for ~20 s and ~10 s before the first two passes of rolling (surface cooling rate of approximately 10–15 °C/s) to reduce the surface temperature to ~820 °C to establish a temperature gradient between the center and the surface. Subsequent passes of rolling do not require water cooling and are continuous. Due to the different resistance to deformation at different locations caused by the temperature gradient, it is beneficial for the deformation of the center during the rolling process.

Table 1. The chemical composition of the experimental steel (wt.%).

C	Si	Mn	Cr	Ni	Cu	Mo	V	P	S
0.069	0.34	0.59	0.64	4.79	0.024	0.5	0.065	0.0053	0.0004

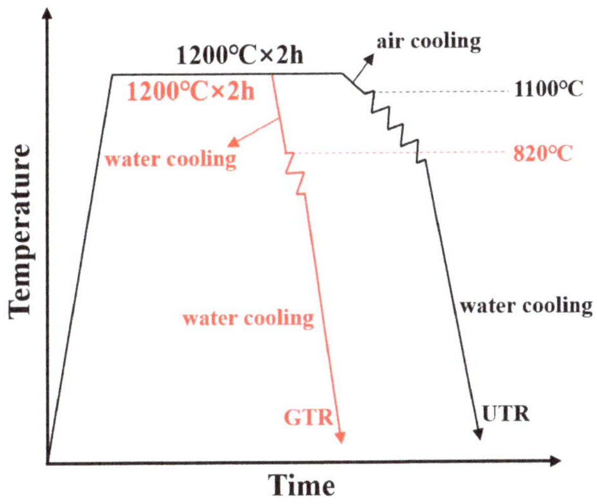

Figure 1. Thermo-mechanical controlled processing (TMCP) process flow chart of experimental steels.

The specimens for optical microscope (OM), scanning electron microscopy (SEM), and electron backscattered diffraction (EBSD) observations were cut along the RD–ND plane (RD: rolling direction, ND: normal direction). They were taken from 1/2 location along the thickness direction and henceforth referred to as UTR-1/2 and GTR-1/2, respectively. Specimens for microstructural studies were observed using a Sigma 300 field emission scanning electron microscopy (FE-SEM) (Sigma, Livonia, MI, USA). EBSD analysis was conducted using TESCAN MIRA 3 LMH FE-SEM equipped (TESCAN MIRA, Dublin, Ireland) with an Oxford Nordlys Nano Max 3 EBSD detector (Oxford Instruments, Oxford, UK) at an acceleration voltage of 20 kV and a step size of 0.9 μm. EBSD data were post-processed by HKL CHANNEL 5 (Version 5.12.72.0, Oxford Instruments, Oxford, UK) flamenco software to acquire the necessary information. The results of EBSD were used to analyze boundary density by a Python program.

Two tensile specimens were tested for each experimental steel (after rolling and water cooling to room temperature) and the average values were taken for the results of tensile tests. Round bar tensile specimens with a 5 mm gage diameter and 25 mm gage length were tested at room temperature at a crosshead speed of 1 mm/min. Three impact specimens with dimensions of 55 mm × 10 mm × 10 mm were tested for each experimental steel at −80 °C and the average values were taken for the results of instrumented Charpy v-notch (CVN) impact tests. Tensile specimens and CVN impact specimens were cut from the plates along the longitudinal direction.

3. Results

Figure 2 shows the OM micrographs and SEM micrographs of different positions of the experimental steels. It can be seen from the figure that for both steel UTR and steel GTR, the microstructure is mainly martensite, with obvious martensite lath. It can be seen from the prior austenite marked by the red dotted line in Figure 2 that the prior austenite grain size of steel GTR is obviously smaller than that of steel UTR and that the martensite lath is finer.

In order to analyze the boundary information more precisely, the EBSD experimental data are analyzed using the HKL CHANNEL 5 flamenco software. The resulting inverse pole figure color map is shown in Figure 3. Different colors represent martensitic crystal units that are aligned with the observation plane in the vertical direction, as shown in the inset. The closer the color is, the smaller the difference in orientation is, and vice versa. It can be seen from Figure 3 that the grain size of steel GTR in the center is significantly smaller, which is mainly due to the microstructural refinement effect brought about by the GTR process, which will be further explained later.

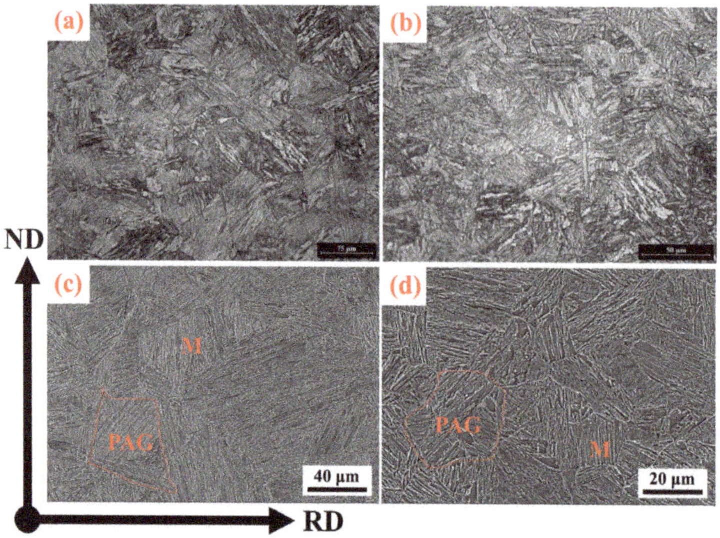

Figure 2. OM micrographs ((**a**) 128×, (**b**) 256×) and SEM micrographs ((**c**) 500×, (**d**) 1000×) of steel UTR and steel GTR in the center. ((**a**): UTR-1/2, (**b**): GTR-1/2, (**c**): UTR-1/2, (**d**): GTR-1/2, RD: Rolling direction, ND: Normal direction).

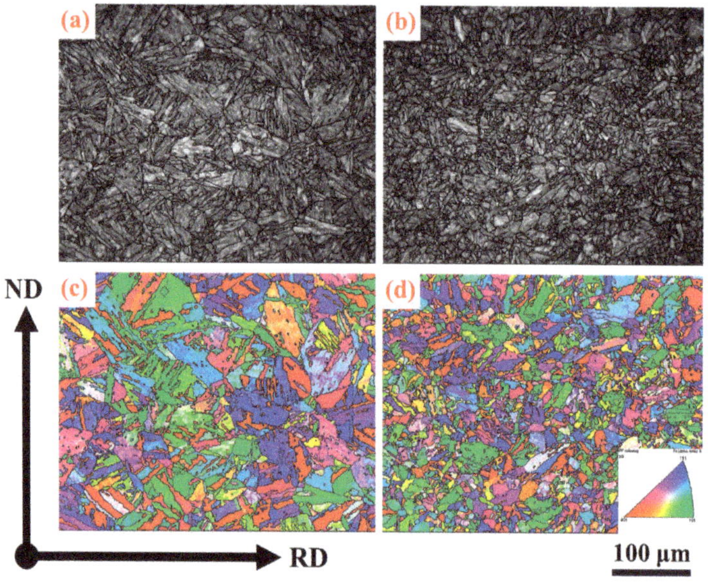

Figure 3. BC (Band contrast) map and IPFX (inverse pole figure *X*-axis) color map of steel UTR and steel GTR in the center. ((**a**): UTR-1/2, (**b**): GTR-1/2, (**c**): UTR-1/2, (**d**): GTR-1/2, RD: Rolling direction, ND: Normal direction).

The results of tensile tests at room temperature and instrumented CVN impact tests at −80 °C are shown in Table 2. It can be seen that the yield strength ratio of steel GTR-1/2 is not much different from steel UTR-1/2, the yield strength of steel GTR-1/2 is ~22 MPa higher than steel UTR-1/2, and the tensile strength of steel GTR-1/2 is ~37 MPa higher. In addition, the plasticity and toughness have been significantly improved, especially the toughness, which is ~71.2 J higher than steel UTR-1/2. These are the differences brought about by different rolling processes; the reasons for the differences will be emphasized later.

Table 2. Mechanical properties of experimental steels.

Number	YS [1] (MPa)	TS [2] (MPa)	Yield Ratio	El [3] (%)	−80 °C CVN [4] (J)
UTR-1/2	769.5	1083.5	0.71	15.6	168.8
GTR-1/2	791.5	1120.5	0.71	17.7	240.0

[1] Yield strength; [2] Tensile strength; [3] Elongation; [4] Charpy v-notch impact toughness.

Figure 4 is the SEM images of the fracture surface of impact specimens. It can be seen from the figure that the shear lip area of steel GTR-1/2 is larger than steel UTR-1/2. From the enlarged fracture surface, it can be seen that the fracture surface of steel UTR-1/2 is mainly composed of cleavage planes, shear cracks, and a few dimples, showing a clear brittle fracture. The fracture surface of steel GTR-1/2 is mainly composed of large and small dimples, with the presence of dimples with tearing ridges, indicating plastic deformation during the crack propagation process and high crack propagation resistance. The crack propagation requires a large amount of energy, showing a clear ductile fracture [14,15]. The characteristics of the fracture surface also confirm the difference in low-temperature toughness between steel UTR-1/2 and steel GTR-1/2. The GTR process can significantly increase the proportion of dimples and improve the low-temperature toughness.

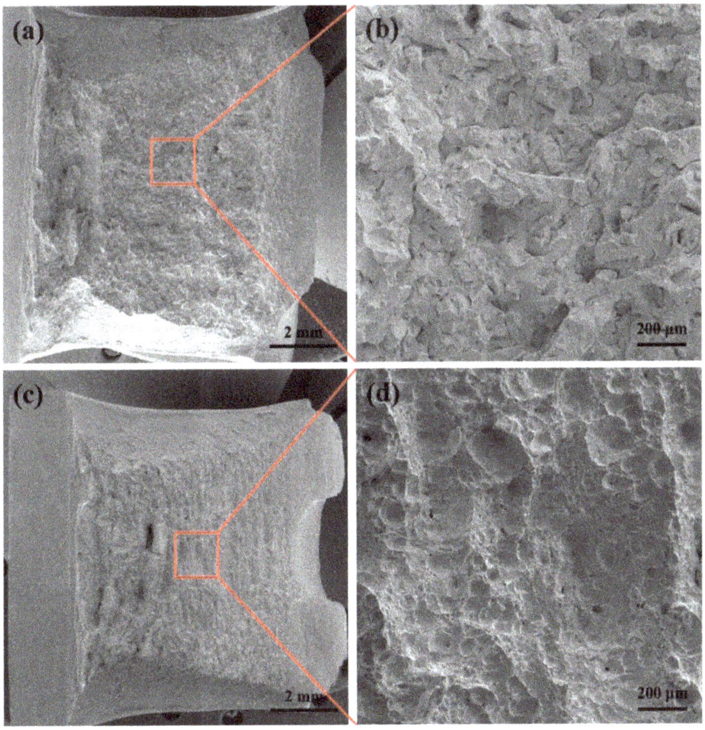

Figure 4. SEM appearance of fractures of the UTR-1/2 ((**a**) 27×, (**b**) 200×)) and GTR-1/2 ((**c**) 27×, (**d**) 200×).

4. Discussion

4.1. Research on Strength

Figure 5 shows the prior austenite grain boundaries distribution map and various grain boundary density statistical diagrams calculated by the software [16]. Through the statistics of the prior austenite grain size in Figure 5a,b, the average prior austenite grain size of steel UTR-1/2 and steel GTR-1/2 is 41.1 ± 6.6 μm and 20.9 ± 3.4 μm, respectively. Compared with the UTR process, the GTR process significantly reduces the prior austenite grain size, which is mainly due to the reduction in the surface temperature to ~820 °C before rolling, forming a temperature gradient from low to high from the surface to the center, increasing the strain and accumulative deformation in the center, and refining the prior austenite grain size in the center. The refinement of austenite grain increases the crystal defects such as vacancies, dislocations, deformation bands, etc., inside the austenite. This causes an increase in non-uniform nucleation points, which is beneficial to grain refinement after transformation [17–20]. Meanwhile, the research shows that the refinement of the prior austenite grain is conducive to the increase in the density of the packet boundaries and block boundaries in martensite [15,21]. As can be seen from Figure 5c, compared to the UTR process, the GTR process does not significantly increase the density of block boundaries and sub-block boundaries in the center but significantly increases the density of PAG boundaries, high-packet boundaries, and low-packet boundaries. Compared to steel UTR-1/2, the high-packet boundaries density of steel GTR-1/2 increases from 0.075 μm^{-1} to 0.104 μm^{-1}, the low-packet boundaries density increases from 0.037 μm^{-1} to 0.055 μm^{-1}, and the PAG boundaries density increases from 0.055 μm^{-1} to 0.132 μm^{-1}.

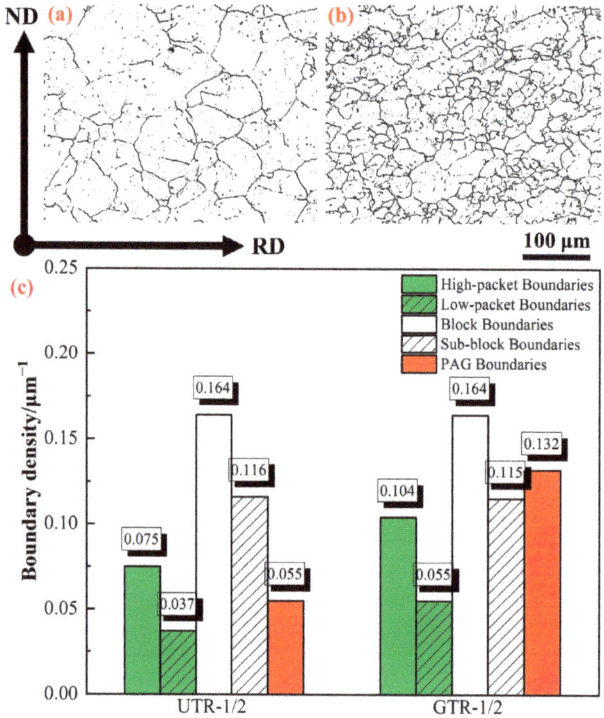

Figure 5. (**a**,**b**) PAG (prior austenite grain) boundaries skeleton map; (**c**) Boundary density computed by the Python program (PAG boundaries-prior austenite grain boundaries; High-packet boundaries (misorientation angle ≥ 15°); Low-packet boundaries (misorientation angle < 15°)). (**a**) and (**b**) are maps for UTR-1/2 and GTR-1/2, respectively.

In addition to grain boundaries, dislocations are also an important factor affecting strength. The dislocation density can be estimated using GND (geometrically necessary dislocation) density. The grain boundary distribution and GND density of the experimental steels are shown in Figure 6. The closer the color is to red, the higher the dislocation density is, and the closer the color is to blue, the lower the dislocation density is. Some studies have shown that GND can be calculated by kernel average misorientation (KAM) and the dislocation density can be estimated by calculating GND density using Equation (1) [22,23]. The calculation results are shown in Table 3.

$$\rho = \frac{2\theta}{ub} \tag{1}$$

where θ is the misorientation angle, which is directly obtained from EBSD data and is less than 2°, u is the unit length (step size = 0.9 μm), and b is the Burgers vector and is equal to 0.248 nm.

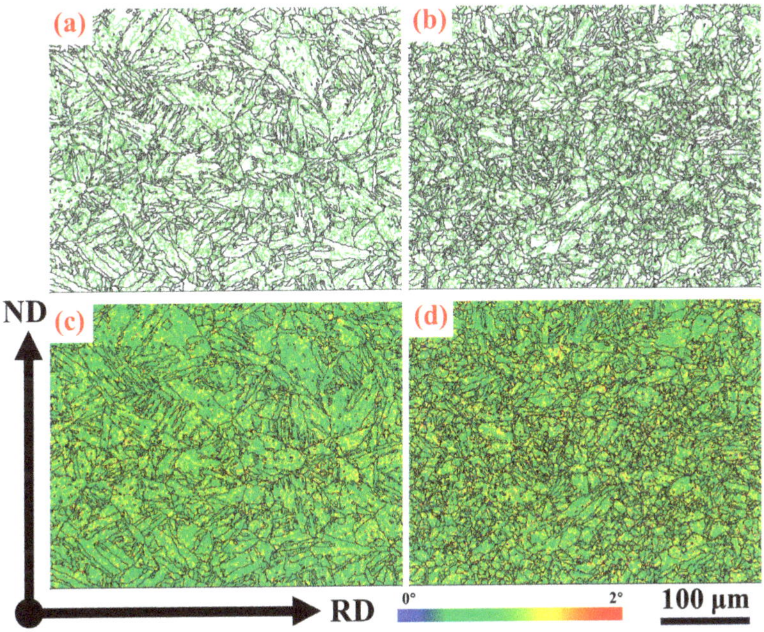

Figure 6. (**a**,**b**) Distribution map of high-angle grain boundaries (misorientation angle ≥ 15°) and low-angle grain boundaries (misorientation angle < 15°) (black lines: high-angle grain boundaries, green lines: low-angle grain boundaries); (**c**,**d**) GND (geometrically necessary dislocation) density maps. (**a**,**c**) and (**b**,**d**) are maps for UTR-1/2 and GTR-1/2, respectively.

Table 3. The density of grain boundaries and dislocations.

Number	Dislocation Density (ρ)/m^{-2}	$\Delta\sigma_{Dis}$/MPa
UTR-1/2	1.42×10^{14}	283.9
GTR-1/2	1.51×10^{14}	292.7

As can be seen from Table 3, there is a difference in dislocation density between steel GTR-1/2 and steel UTR-1/2 and the strength contribution of dislocation strengthening is calculated using Equation (2) [24].

$$\Delta\upsilon_{Dis} = \alpha MGb\sqrt{\rho} \tag{2}$$

where α is a constant with value of 0.435 [25], M is the Taylor factor and for ferritic steel is 2.75 [26], G is the shear modulus of 80,300 MPa [25], b is the Burgers vector of 0.248 nm, and ρ is the dislocation density (Table 3). The calculation results of dislocation strengthening contributions are shown in Table 3. The strength contribution of dislocation strengthening for steel GTR-1/2 is 292.7 MPa and the strength contribution of dislocation strengthening for steel UTR-1/2 is 283.9 MPa, with a small gap between the two. However, from the strength data in Table 2, it can be seen that there is still a larger gap in the yield strength between the two, indicating that in addition to dislocation strengthening, other strengthening methods also have a certain influence. For the same composition and with less precipitation, grain boundary strengthening is the only possible strengthening method that can affect the strength.

From Figure 6a,b, it can be seen that there is a slight difference in the density of high-angle and low-angle grain boundaries between steel GTR-1/2 and steel UTR-1/2. The strength contribution estimation based on the density of high-angle and low-angle grain boundaries is limited, as not all grain boundaries have a significant impact on strength. Therefore, refined analysis of grain boundaries is necessary. Some studies have shown that both block boundaries and sub-block boundaries can hinder dislocation motion and significantly improve strength [27]. At the same time, some studies have shown that block boundaries are the grain boundary factor with the highest impact on strength [28]. For steel GTR-1/2 and steel UTR-1/2, the densities of block boundaries and sub-block boundaries are basically the same, indicating that other grain boundary types also have a certain contribution to strength.

Some studies have used the slip transmission factor m' as a geometric factor to measure the ability of adjacent grains to slip transfer without loss of the coherence at the grain boundary, calculated by $m' = \cos(\phi)\cos(\kappa)$, where ϕ is the angle between the slip directions in the neighboring grains and κ is the angle between normal to the slip planes of neighboring grains [29,30]. At the same time, the slip transmission coefficient between different variants is related to plasticity and strength [30,31]. According to the relevant research method, the value of N (the maximum m' of all slip systems in the adjacent sub-volumes) is used to represent the difficulty of slip transmission of a grain boundary, so the larger the value of 1-N, the stronger the resistance to slip transmission [31]. Based on phase transformation crystallography calculations, Packet boundaries, especially high-angle packet boundaries, have higher 1-N values compared to block boundaries or sub-block boundaries. The orientation of PAG approximates a random orientation distribution; hence, they also possess higher 1-N values [32]. The statistical results of the 1-N values for actual specimens are shown in Figure 7. As can be seen from the figure that the block boundary accounts for a higher proportion in steel GTR-1/2 compared to steel UTR-1/2 but due to its lower 1-N value, it has little effect on improving the anti-slip ability of the grain boundary and is not the main reason for the difference in strength. Instead, the local enlarged image shows that there are significant differences between the PAG boundaries and the high-angle packet boundaries, especially the PAG boundaries. That is mainly because the GTR process significantly increases the deformation penetration in the center to refine the prior austenite grains. Furthermore, the refinement of the austenite grains also affects the tendency for variant selection, resulting in the formation of more Packet boundaries. The increased density of PAG boundaries and Packet boundaries with higher 1-N values implies that the effective grains contributing to strength enhancement have been refined. Therefore, it can significantly improve the strength of the center.

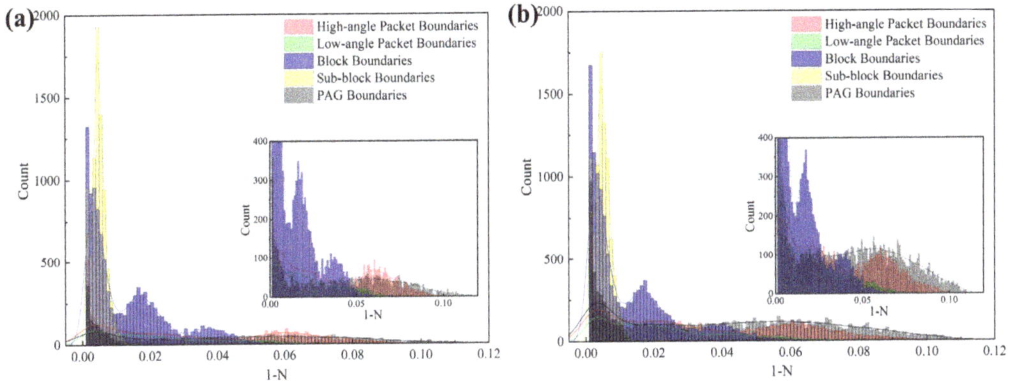

Figure 7. The 1-N value of different types of grain boundaries distribution map (the small image is a locally enlarged image): (**a**) UTR-1/2; (**b**) GTR-1/2.

4.2. Research on Toughness

The impact load/energy displacement curve obtained from the instrumented Charpy impact test is shown in Figure 8. In the figure, F_m is the starting point of stable crack propagation, before which the crack initiation is stable and the energy consumed before F_m is the crack initiation energy (CIE). F_{iu} is the starting point of unstable crack propagation. F_a is the starting point of plastic fracture, where the unstable fracture is suppressed and enters the plastic fracture stage. The plastic fracture front is formed between the F_{gy} (general yield force) and F_m. The initial expansion of the crack is stable until unstable expansion occurs and the expansion then stops in the suppression stage [33]. During this period, the energy absorbed by the impact specimen is crack propagation energy (CPE). During stable crack propagation, plastic deformation and cleavage crack propagation occur simultaneously but once unstable propagation begins, the crack will rapidly extend and lead to specimen failure within a short time. Therefore, the impact absorption energy during the unstable crack propagation stage is much lower than during the stable propagation stage. The impact load/displacement curve of steel UTR-1/2 corresponds to the "E" type in the standard curve, with plastic deformation before F_m followed by stable and unstable crack propagation [34]. The impact load/displacement curve of steel GTR-1/2 corresponds to the "F" type in the standard curve, with plastic deformation before F_m followed only by stable crack propagation [34]. Compared to steel GTR-1/2, steel UTR-1/2 has a smaller elastic deformation work, while the plastic deformation work is essentially the same. After reaching the starting point of stable crack propagation, steel UTR-1/2 undergoes a certain degree of stable crack propagation, while when the crack propagates to a certain size, it undergoes unstable crack propagation, exhibiting semi-brittleness. The crack of steel GTR-1/2 stably propagated to fracture without instability, showing good low-temperature toughness.

The overall misorientation angle (OMA) is often used as a standard to define high-angle grain boundaries and low-angle grain boundaries. The specific misorientation angle (SMA) is used to define the misorientation angle between specific crystallographic planes. The [28] planes are the primary slip plane, so it is believed that the {110}-SMA controls plastic deformation and grain boundaries with high {110}-SMA can be effective for the free slip [31]. This study statistically analyzes the density of {110}-SMA as a basis for analyzing toughness differences and the statistical results are shown in Figure 9. From the figure, it can be seen that block boundaries and sub-block boundaries are mainly concentrated in areas with lower {110}-SMA and the difference between the two is not significant. Although there are some differences in the density of low-angle packet boundaries, they are mainly concentrated in the lower {110}-SMA. The effect of these three types of grain boundaries on hindering dislocation gliding is not obvious. However, the density of high-angle packet boundaries and PAG boundaries in steel GTR-1/2 are significantly higher than steel UTR-1/2 and mainly concentrated in areas with higher {110}-SMA. Therefore, these two types of grain boundaries can effectively hinder dislocation gliding, thereby enhancing the plastic deformation capability of the steel plate. Although the GTR process and the UTR process lead to similar CIE at −80 °C, the differences in CPE reflect the differences in their plastic deformation capabilities. Due to the presence of a plastic zone at the crack tip, during stable crack propagation, boundaries with higher {110}-SMA can alleviate stress concentration, allowing for the absorption of more impact energy. That reduces the propensity for unstable crack propagation, increasing impact absorption capacity. The absence of unstable crack propagation and the significantly higher CPE in GTR-1/2 steel compared to UTR-1/2 can be attributed to the higher density of high-angle Packet boundaries and PAG boundaries. Therefore, the GTR process can significantly refine the size of PAG in the center compared to the UTR process. Meanwhile, they increase the density of {110}-SMA in the PAG boundaries and high-angle packet boundaries, playing a role in alleviating stress concentration at the crack tip and thus improving the stable crack propagation, improving low-temperature toughness.

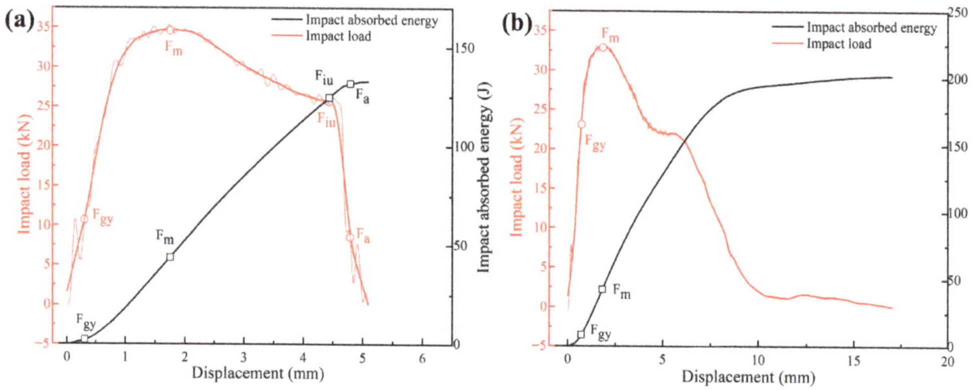

Figure 8. CVN impact load and absorbed energy vs. displacement curves obtained in the instrumented Charpy impact tests conducted at −80 °C: (**a**) UTR-1/2; (**b**) GTR-1/2.

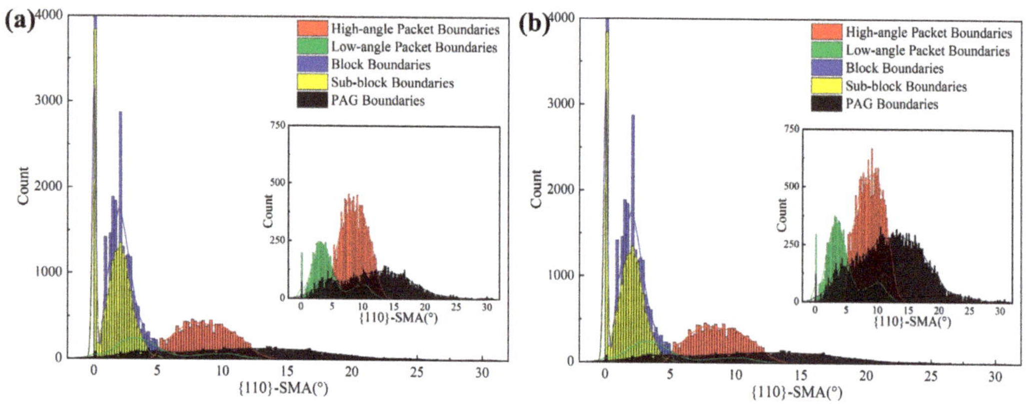

Figure 9. {110}-SMAs of different types of grain boundaries (the small image is a locally enlarged image): (**a**) UTR-1/2; (**b**) GTR-1/2.

5. Conclusions

Compared to the traditional UTR process, the GTR process can significantly refine the size of the PAG in the center while improving the strength, plasticity, and low-temperature toughness of the center. The GTR process not only increases the PAG boundaries density in the center but also significantly increases the packet boundaries density. Through the analysis of dislocation density and grain boundary density, it is shown that the GTR process contributes more to the dislocation strengthening of the center and the proportion of PAG boundaries and high-angle packet boundaries with larger 1-N values is higher, resulting in stronger resistance to slip transfer. Through instrumented Charpy impact test and grain boundary density analysis, it is shown that the GTR process has a higher proportion of high {110}-SMA of the PAG boundaries and high-angle packet boundaries in the center, effectively hindering dislocation gliding, increasing the stable crack propagation, and improving low-temperature toughness.

Author Contributions: Conceptualization, J.C. and X.W.; methodology, J.C.; software, J.Z.; validation, J.C. and X.W.; formal analysis, J.C.; investigation, Z.W.; resources, X.W.; data curation, J.C.; writing—original draft preparation, J.C.; writing—review and editing, J.C.; visualization, J.Z. and Z.W.; supervision, X.W.; project administration, X.W.; funding acquisition, X.W. All authors have read and agreed to the published version of the manuscript.

Funding: This research was funded by the Fundamental Research Business Funds for Central Universities (FRF-BD-22-02).

Data Availability Statement: The raw data supporting the conclusions of this article will be made available by the authors on request.

Acknowledgments: The authors are grateful to Hongyu Wu, Yu Li, and Zhaowan Xie for their help with writing and English editing.

Conflicts of Interest: Author Zhongwen Wu was employed by the company Xiangtan Iron & Steel Group Co., Ltd. The remaining authors declare that the re-search was conducted in the absence of any commercial or financial relationships that could be construed as a potential conflict of interest.

ferences

Zhou, T.; Yu, H.; Wang, S. Effect of microstructural types on toughness and microstructural optimization of ultra-heavy steel plate: EBSD analysis and microscopic fracture mechanism. *Mater. Sci. Eng. A* **2016**, *658*, 150–158. [CrossRef]

Liu, D.; Cheng, B.; Chen, Y. Strengthening and Toughening of a Heavy Plate Steel for Shipbuilding with Yield Strength of Approximately 690 MPa. *Metall. Mater. Trans. A* **2012**, *44*, 440–455. [CrossRef]

Zhou, T.; Yu, H.; Wang, S. Microstructural Characterization and Mechanical Properties across Thickness of Ultra-Heavy Steel Plate. *Steel Res. Int.* **2017**, *88*, 1700132. [CrossRef]

Xie, B.-S.; Cai, Q.-W.; Yun, Y.; Li, G.-S.; Ning, Z. Development of high strength ultra-heavy plate processed with gradient temperature rolling, intercritical quenching and tempering. *Mater. Sci. Eng. A* **2017**, *680*, 454–468. [CrossRef]

Guo, K.; Pan, T.; Zhang, N.; Meng, L.; Luo, X.; Chai, F. Effect of Microstructural Evolution on the Mechanical Properties of Ni-Cr-Mo Ultra-Heavy Steel Plate. *Materials* **2023**, *16*, 1607. [CrossRef] [PubMed]

Wang, Q.; Ye, Q.; Tian, Y.; Fu, T.; Wang, Z. Superior Through-Thickness Homogeneity of Microstructure and Mechanical Properties of Ultraheavy Steel Plate by Advanced Casting and Quenching Technologies. *Steel Res. Int.* **2021**, *92*, 2000698. [CrossRef]

Zhang, S.; Hu, X.; Du, Y.; Jiang, H.; Pang, H.; Rong, L. Cross-section effect of Ni-Cr-Mo-B ultra-heavy steel plate for offshore platform. *Acta Met. Sin.* **2020**, *56*, 1227–1238.

Wei, Y.; Gaosheng, L.; Qingwu, C. Effect of a novel gradient temperature rolling process on deformation, microstructure and mechanical properties of ultra-heavy plate. *J. Mater. Process. Technol.* **2015**, *217*, 317–326. [CrossRef]

Tao, P.; Yu, H.; Fan, Y.; Fu, Y. Effects of cooling method after intercritical heat treatment on microstructural characteristics and mechanical properties of as-cast high-strength low-alloy steel. *Mater. Des. (1980–2015)* **2014**, *54*, 914–923. [CrossRef]

Tsuyama, S. Thick Plate Technology for the Last 100 Years: A World Leader in Thermo Mechanical Control Process. *ISIJ Int.* **2015**, *55*, 67–78. [CrossRef]

Nishioka, K.; Ichikawa, K. Progress in thermomechanical control of steel plates and their commercialization. *Sci. Technol. Adv. Mater.* **2012**, *13*, 023001. [CrossRef] [PubMed]

Omata, K.; Yoshimura, H.; Yamamoto, S. Leading high performance steel plates with advanced manufacturing technologies. *NKK Technol. Rev.* **2003**, *88*, 8.

13. Gaosheng, L.; Wei, Y.; Qingwu, C. Investigation of the Evolution of Central Defects in Ultra-Heavy Plate Rolled Using Gradie Temperature Process. *Metall. Mater. Trans. B* **2015**, *46*, 831–840. [CrossRef]
14. Kang, J.; Wang, C.; Wang, G.D. Microstructural characteristics and impact fracture behavior of a high-strength low-alloy ste treated by intercritical heat treatment. *Mater. Sci. Eng. A* **2012**, *553*, 96–104. [CrossRef]
15. Li, X.; Lu, G.; Wang, Q.; Zhao, J.; Xie, Z.; Misra RD, K.; Shang, C. The Effects of Prior Austenite Grain Refinement on Strength ar Toughness of High-Strength Low-Alloy Steel. *Metals* **2021**, *12*, 28. [CrossRef]
16. Li, X.C.; Zhao, J.X.; Cong, J.H.; Misra RD, K.; Wang, X.M.; Wang, X.L.; Shang, C.J. Machine learning guided automatic recogniti of crystal boundaries in bainitic/martensitic alloy and relationship between boundary types and ductile-to-brittle transiti behavior. *J. Mater. Sci. Technol.* **2021**, *84*, 49–58. [CrossRef]
17. Cong, J.; Li, J.; Fan, J.; Misra, R.D.K.; Xu, X.; Wang, X. Effect of austenitic state before ferrite transformation on the mechanical behavi at an elevated temperature for seismic-resistant and fire-resistant constructional steel. *J. Mater. Res. Technol.* **2021**, *13*, 1220–12 [CrossRef]
18. Han, P.; Liu, Z.P.; Li, Q.; Xie, Z.J.; Wang, X.L.; Misra, R.D.K.; Shang, C.J. A phenomenological understanding of the novel desi of hierarchical structure for 1 GPa ultrahigh strength and high toughness combination low alloy steel. *Mater. Sci. Eng. A* **20** *881*, 145387. [CrossRef]
19. Xie, Z.J.; Han, P.; Liu, Z.P.; Wang, X.L.; Shang, C.J. Influence of initial microstructure on reaustenitization behavior in low all steel by in-situ high-temperature EBSD characterization. *Mater. Lett.* **2023**, *350*, 134876. [CrossRef]
20. Han, P.; Liu, Z.; Xie, Z.; Wang, H.; Jin, Y.; Wang, X.; Shang, C. Influence of band microstructure on carbide precipitation behavi and toughness of 1 GPa-grade ultra-heavy gauge low-alloy steel. *Int. J. Miner. Metall. Mater.* **2023**, *30*, 1329–1337. [CrossRef]
21. Furuhara, T.; Kikumoto, K.; Saito, H.; Sekine, T.; Ogawa, T.; Morito, S.; Maki, T. Phase Transformation from Fine-grained Austeni *Isij Int.* **2008**, *48*, 1038–1045. [CrossRef]
22. Kubin, L.P.; Mortensen, A. Geometrically necessary dislocations and strain-gradient plasticity: A few critical issues. *Scr. Mat* **2003**, *48*, 119–125. [CrossRef]
23. Calcagnotto, M.; Ponge, D.; Demir, E.; Raabe, D. Orientation gradients and geometrically necessary dislocations in ultrafi grained dual-phase steels studied by 2D and 3D EBSD. *Mater. Sci. Eng. A* **2010**, *527*, 2738–2746. [CrossRef]
24. Cong, J.; Li, J.; Fan, J.; Liu, P.; Misra, R.D.K.; Shang, C.; Wang, X. The Impact of Interphase Precipitation on the Mechanic Behavior of Fire-Resistant Steels at an Elevated Temperature. *Materials* **2020**, *13*, 4294. [CrossRef] [PubMed]
25. Gladman, T. *The Physical Metallurgy of Microalloyed Steels*; The Institute of Materials: London, UK, 1997.
26. Sun, X.; Choi, K.S.; Soulami, A.; Liu, W.N.; Khaleel, M.A. On key factors influencing ductile fractures of dual phase (DP) stee *Mater. Sci. Eng. A* **2009**, *526*, 140–149. [CrossRef]
27. Du, C.; Hoefnagels, J.P.M.; Vaes, R.; Geers, M.G.D. Block and sub-block boundary strengthening in lath martensite. *Scr. Mat* **2016**, *116*, 117–121. [CrossRef]
28. Wang, J.; Hong, H.; Huang, A.; Yang, X.; Qian, R.; Shang, C. New insight into the relationship between grain boundaries ar hardness in bainitic/martensitic steels from the crystallographic perspective. *Mater. Lett.* **2022**, *308*, 131105. [CrossRef]
29. Weaver, J.S.; Nan, L.; Mara, N.A.; Jones, D.R.; Hansohl, C.; Bronkhorst, C.A.; Fensin, S.J.; Gray, G.T. Slip transmission of hig angle grain boundaries in body-centered cubic metals: Micropillar compression of pure Ta single and bi-crystals. *Acta Mat* **2018**, *156*, 356–368. [CrossRef]
30. Wang, J.; Qian, R.; Huang, S.; Shang, C. Effect of Double-Quenching on the Hardness and Toughness of a Wear-Resistant Ste *Metals* **2022**, *13*, 61. [CrossRef]
31. Wang, Z.; Guo, Z.; Shang, C.; Chen, B.; Hui, Y. Characterization of the Stretch Flangeabitity of High-Strength Bainitic Steel: T Significance of Variant Pairs. *Materials* **2022**, *15*, 276. [CrossRef]
32. Zhao, J.; Li, X.; Wang, X.; Liu, S.; Wang, X.; Shang, C. Distribution feature of specific misorientation angle in a bainitic steel. *Mat Charact.* **2021**, *172*, 110874. [CrossRef]
33. Chaouadi, R.; Fabry, A. On the utilization of the instrumented Charpy impact test for characterizing the flow and fractu behavior of reactor pressure vessel steels. In *European Structural Integrity Society*; Elsevier: Amsterdam, The Netherlands, 20C Volume 30, pp. 103–117.
34. *ISO 14556:2015*; Metallic Materials—Charpy V-Notch Pendulum Impact Test—Instrumented Test Method. ES-AENOR: Madri Spain, 2016; Volume UNE-EN ISO 14556-2016.

Disclaimer/Publisher's Note: The statements, opinions and data contained in all publications are solely those of the individu author(s) and contributor(s) and not of MDPI and/or the editor(s). MDPI and/or the editor(s) disclaim responsibility for any injury people or property resulting from any ideas, methods, instructions or products referred to in the content.

metals

icle

:udy on the Effect of Niobium on the High Temperature xidation Resistance of Ferritic Stainless Steel

Zhu [1], Dianxiu Xia [1,*], Peidun Chen [2], Qing Han [1], Xuelin Wang [3], Zhiheng Liu [1] and Kun Jiang [1]

1 School of Mechanical Engineering, University of Jinan, Jinan 250022, China; 202121100309@stu.ujn.edu.cn (E.Z.); me_hanq@ujn.edu.cn (Q.H.); 202121100308@stu.ujn.edu.cn (Z.L.); 202221201020@stu.ujn.edu.cn (K.J.)
2 Shandong Taishan Iron and Steel Group Co., Ltd., Jinan 271100, China; tgcpd@126.com
3 School of Materials Science and Engineering, University of Science and Technology, Beijing 100083, China; xuelin2076@ustb.edu.cn
* Correspondence: me_xiadx@ujn.edu.cn

Abstract: More and more cars are installing urea selective catalytic reduction (SCR) systems to solve the problem of exhaust emissions, which often operates at high temperatures in the exhaust system and is prone to failure of the exhaust pipe due to high-temperature oxidation. And niobium containing ferritic stainless steel has been widely used in the manufacturing of automotive exhaust pipes. In order to extend the service life of ferrite stainless steel exhaust pipes, niobium plays a crucial role as an added alloying element. The solid solution and precipitation of niobium in ferritic stainless steel will give ferritic stainless steel more excellent high temperature resistance. The precipitation of Nb can change the organizational structure in steel and refine the grains. However, if the content is not properly controlled, large particles of (Nb, Ti) C will precipitate, which will reduce the high temperature oxidation resistance. In this paper, the high temperature oxidation behavior of two kinds of ferritic stainless steels with different Nb content at 700 °C, 800 °C and 900 °C was studied. The microstructure of the oxide film on the surface of the material, the thickness of the oxide layer on the cross section, the distribution of chemical composition, the existence form and distribution of Nb element were analyzed by SEM, EDS, XRD and TEM. The results show that the higher the niobium content, the better the high temperature oxidation resistance, and the higher the temperature, the more obvious the high temperature oxidation resistance of niobium. This is because the high Nb content steel is easy to precipitate NbN and other Nb-containing precipitates at the grain boundary, which is helpful to the high temperature resistance. In the case of less Nb content, its ability to generate (Nb, Ti) C precipitates and coarsen at high temperatures reduces its high-temperature resistance.

Keywords: ferritic stainless steel; niobium; high-temperature oxidation; oxide film peeling; material failure; chromium diffusion

1. Introduction

In order to reduce pollution from automobile exhausts to the atmosphere, various countries have been continuously improving automobile exhaust emission standards in recent years. In order to meet the strict requirements of automobile exhaust emission standards, more and more commercial diesel vehicles have installed urea-selective catalytic reduction (SCR) exhaust aftertreatment devices. After installing the SCR exhaust aftertreatment device, the operating temperature of the automobile exhaust system is significantly increased, up to 900 °C, and more complications such as thermal expansion and contraction, thermal cycle and ammonia nitrogen atmosphere corrosion occur. The problem of exhaust system failure due to high-temperature oxidation is becoming more and more serious [1].

Previous studies have investigated nitrogen alloy austenitic stainless steel, which has increased mechanical properties (toughness, and tensile properties) and is used as a component material for automotive exhaust pipes [2]. Niobium-containing ferritic stainless steel also has advantages such as high thermal conductivity, low thermal expansion coefficient, strong corrosion resistance, and lower manufacturing cost compared to austenitic stainless steel. Therefore, it is increasingly used as a raw material for exhaust pipes in SCR systems. Many researchers from China and other countries have studied the high-temperature oxidation behavior of ferritic stainless steel. Studies have shown that stainless steel often has better high-temperature stability after adding niobium [3]. It was found that the solid solution of Nb endows ferritic stainless steel with better high-temperature resistance, and the precipitation of Nb can change the solidification structure and refine the grains. But if the Nb content is not properly controlled, large particles of TiN and (Nb, Ti) (C, N) precipitate, reducing high-temperature performance. High-temperature oxidation failure is an important reason for the reduced service life or even failure and fracture of automotive exhaust system components equipped with SCR devices, leading to significant waste of resources and economic losses. Therefore, in-depth research on the effect of niobium on the high-temperature oxidation resistance of ferritic stainless steel is of great significance for improving the service life of automotive exhaust system components.

However, there is still a lack of in-depth research and more experimental data on the influence of niobium content on the high-temperature properties of niobium-containing ferritic stainless steel during high-temperature oxidation are needed. For example, Tianlong Liu et al. [4] studied 15CrNbTi ferritic stainless steel. The results show that there are three kinds of precipitated phases: (Nb, Ti) C phase, Fe_2Nb and Fe_3Nb_3 C; however, the effect of precipitated phases was not analyzed. Meng Qian et al. [5] conducted oxidation experiments at 800–1100 °C on two types of 310 s heat-resistant steels with different grain sizes and compositions. The results show that the smaller average grain size and more uniform grain size increase the diffusion flux of preferential oxidation elements and reduce the non-uniform growth of oxidation film. Mengqi Zhang et al. [6] and Badin et al. [7] studied the high-temperature oxidation behavior of ferritic stainless steel and found that its oxide layer consists of a small amount of SiO_2 and continuous dense Cr_2O_3, $FeCr_2O_4$, and Cr-Mn spinel oxides, which have high high-temperature oxidation resistance, but the oxidation mechanism has not been analyzed. Abdolvahid Movahedi-Rad et al. [8] studied the high-temperature oxidation behavior of 409 ferritic stainless steel. They believe that the Cr content of ferritic stainless steel is higher than that of austenitic stainless steel, and there is an internal oxidation zone rich in Cr in the oxidation layer. When sufficient Cr is provided, the penetration degree of the reaction front edge is higher.

This paper investigates the influence of niobium on high-temperature oxidation behavior, compares and studies the high-temperature oxidation behavior of two types of niobium-containing ferritic stainless steel, explores the effect of niobium content on the high-temperature oxidation resistance of ferritic stainless steel, and attempts to reveal the mechanism of niobium's effect on the high-temperature oxidation resistance of stainless steel.

2. Materials and Methods

2.1. Materials

The experimental materials were selected from two grades of ferrite stainless steel plates, 439 and 441, with a thickness of 3 mm (as shown in Table 1). For each experiment, 5 samples for each type of stainless steel were cut into blocks, then their surfaces were polished with 400 #, 800 #, 1200 #, 1500 #, and 2000 # sandpaper in sequence. Finally, they were mechanically polished to SPI (A2) polish grade.

Table 1. Main chemical composition of ferrite stainless steel (wt%).

	C	Si	Mn	P	Cr	Ni	Al	Nb	Ti	N
439	0.010	0.37	0.13	0.015	17.59	0.10	0.065	0.17	0.20	0.0084
441	0.010	0.39	0.23	0.016	18.00	0.10	0.037	0.41	0.16	0.0085

2.2. Methods

This experiment was conducted in the atmospheric environment of the laboratory. Three high-temperature oxidation temperatures were used: 700 °C, 800 °C, and 900 °C. An induction heater was used to heat the samples to temperature. They were kept warm for 30 min, and then cooled for 10 min to room temperature to form one cycle. Each type of sample was heated 30 times. After high-temperature oxidation, the sample was ultrasonically cleaned and stored in a drying dish, and then the composition of the oxide film was analyzed using X-ray diffraction (XRD). The microstructure of the oxide layer of the sample was analyzed using scanning electron microscopy (SEM), and its microstructure was analyzed using energy dispersive spectroscopy (EDS). Then the high-temperature oxidized samples were cold cut, and the surface of the samples in the cross-sectional direction was ground and polished to the same extent. The morphology, composition, and precipitation phase distribution of the oxide film and matrix at the cross-section was analyzed using SEM and XRD, and the precipitation of micro particles in the structure was observed using a transmission electron microscope (TEM). Typical positions of the samples were used for characterization, and the test results have generalizability.

3. Results

3.1. Surface High-Temperature Oxidation Morphology of Experimental Materials at Different Temperatures

The macroscopic and SEM morphology of the sample surface after oxidation of two types of stainless steel at different oxidation temperatures are shown in Figures 1–3. SEM figures with typical features were selected from 5 samples for each type of stainless steel in the article. Figure 1a,d show the macroscopic morphology of the oxidized surfaces of two materials after oxidation at 700 °C. It can be observed that there is still a significant metallic luster, indicating that there are still unoxidized parts or very thin oxide layers on the surfaces of the two materials at this temperature. When the oxidation temperature increased to 800 °C, the metallic luster on the 439 stainless steel sample disappeared and a complete oxidation layer appeared, as shown in Figure 1b. The oxidation layer on the 441 stainless steel sample is relatively complete, but some metallic luster remains, as shown in Figure 1e. As the oxidation temperature continued to rise to 900 °C, the surfaces of the two samples showed almost no metallic luster, were completely covered by the oxide layer, and the oxide layer turned black, indicating an increase in thickness of the oxide layer, and some highlighted areas indicated detachment of the oxide layer.

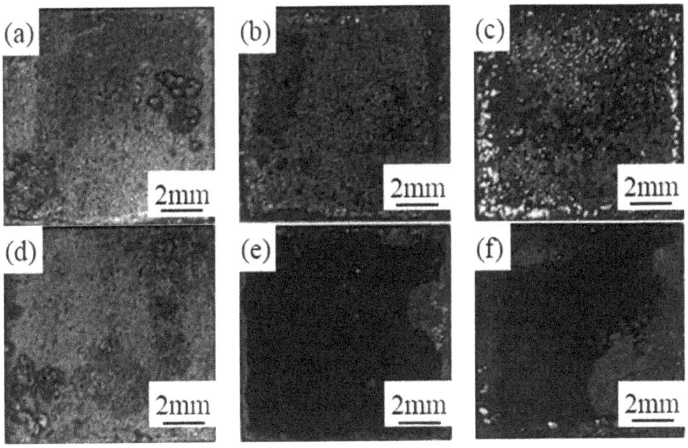

Figure 1. Macromorphology of high-temperature oxidation of samples at different temperatures (**a**) 700 °C (439), (**b**) 800 °C (439), (**c**) 900 °C (439), (**d**) 700 °C (441), (**e**) 800 °C (441), (**f**) 900 °C (441).

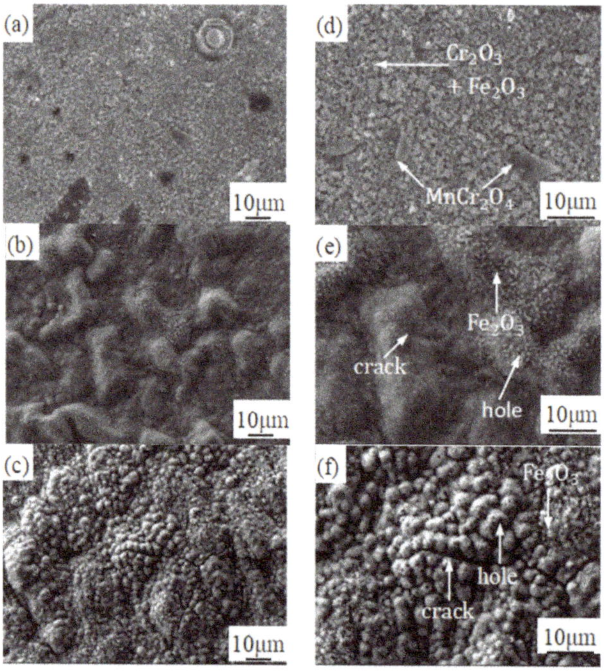

Figure 2. SEM morphology of 439 ferritic stainless steel during high temperature oxidation at different temperatures (**a**,**d**) 700 °C, (**b**,**e**) 800 °C, (**c**,**f**) 900 °C.

Figure 3. SEM morphology of high-temperature oxidation of 441 ferritic stainless steel at different temperatures (**a**,**d**) 700 °C, (**b**,**e**) 800 °C, (**c**,**f**) 900 °C.

Figure 2 shows the surface oxidation microstructure (SEM) of 439 stainless steel samples. From Figure 2a,d, it can be seen that after oxidation at a temperature of 700 °C, the 439 sample's surface forms small and densely packed oxide particles, with dense and layered oxidation products. It can be observed that there are also large spinel-like particles on the surface. As shown in Figure 2b,e, the number of surface granular oxides significantly decreases and accumulates together after oxidation at a temperature of 800 °C. Needle-like structures appear on the oxide shell layer (Figure 2e), which significantly reduces the protective effect of the substrate. In the experiment, thermal cycling generates internal stress, which can lead to surface cracking due to the different levels of hardness of various phases on the surface of the oxide layer. Pores appear on the oxide layer, allowing direct contact with unprotected collectives through the external environment of the pores, accelerating the internal oxidation process [4]. The morphology after oxidation at a temperature of 900 °C is shown in Figure 2c,e. The oxide layer rapidly grows and exhibits oxidation instability. The surface of the oxide tumor is composed of large spherical oxide particles, presenting a convex shape, and the surface of the spherical oxide is filled with holes, accompanied by cracks around it.

Figure 3 shows the surface oxidation microstructure (SEM) of 441 stainless steel sample. It can be seen that after oxidation at a temperature of 700 °C, surface granular oxides mainly aggregate in the form of large particles on a small scale, and gradually grow into small clusters on local surfaces. Most of the substrate passivation films have good integrity (Figure 3a,d). As the oxidation temperature rises to 800 °C, large particle oxides on the surface of the sample begin to grow into layered oxide films that accumulate on the substrate surface (Figure 3b,e). When the oxidation temperature rises to 900 °C, there is a significant change in the surface morphology of the sample, with needle-like oxidation nodules appearing on the surface but no cracking (Figure 3c,e). This is similar to the findings of Jacob et al., which indicate that the smaller the oxide particles, the greater the plasticity of the oxide layer, resulting in stronger adhesion between the oxide layer and the matrix [9].

As the oxidation temperature increases, the surface oxidation of both materials becomes increasingly severe.

3.2. *Analysis of Oxide Layer Thickness of Experimental Materials at Different Temperatures*

In order to further analyze the structure of the oxide layer, SEM characterization was performed on the cross-section of the sample oxide layer, and the thickness of the oxide layer was measured (Figures 4–6). The measured thickness of the oxide layer during characterization represents the thickest visible part of the oxide layer on the sample. The large dispersion of thickness measurements may be due to the different expansion coefficients of the matrix and the oxide during the cooling process, resulting in rupture, shedding or spalling.

From Figure 4, it can be seen that with the increase of temperature, the thickness of the oxide film on the surface of the sample gradually increases. From 800 °C to 900 °C, the oxide layer of stainless steel increases greatly, which generally conforms to the Arrhenius relationship [10]. The thickness of the oxide layer of 439 stainless steel is greater than that of 441 stainless steel at three temperatures, especially at 900 °C, and the error bar in Figure 5 shows that the thickness inhomogeneity of the oxide layer on 439 stainless steel is also greater than that of 441 stainless steel. The data show that the thickness of the oxide layer is not much different between the two at 700 °C. But when oxidized at 800 or 900 °C, the thickness of the oxide layer of 439 stainless steel is almost twice that of 441 stainless steel, indicating that the higher the temperature, the more obvious the high-temperature oxidation resistance of niobium [11].

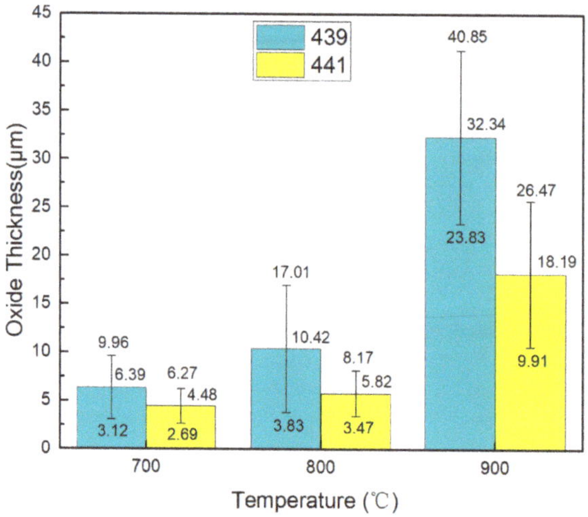

Figure 4. Thickness of two types of ferrite oxide layers at different temperatures.

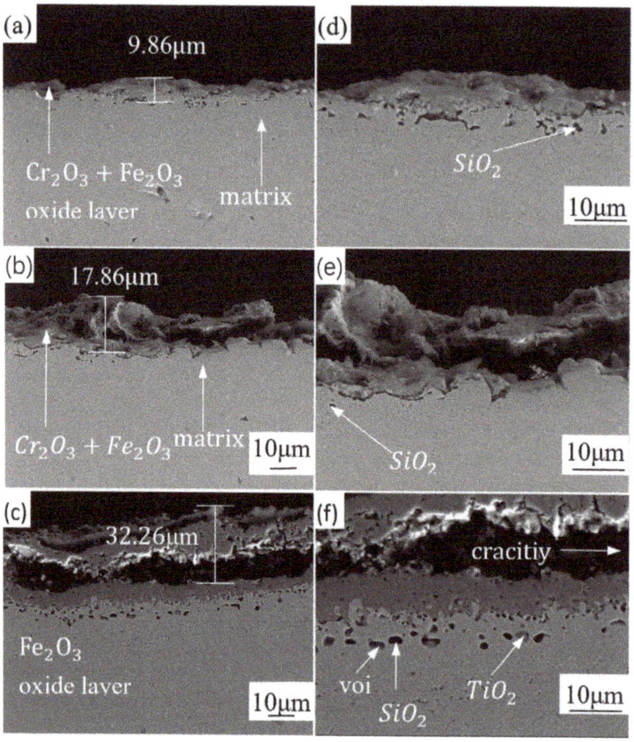

Figure 5. SEM morphology of 439 stainless steel at the interface between substrate and oxide layers at different temperatures (**a**,**d**) 700 °C, (**b**,**e**) 800 °C, (**c**,**f**) 900 °C.

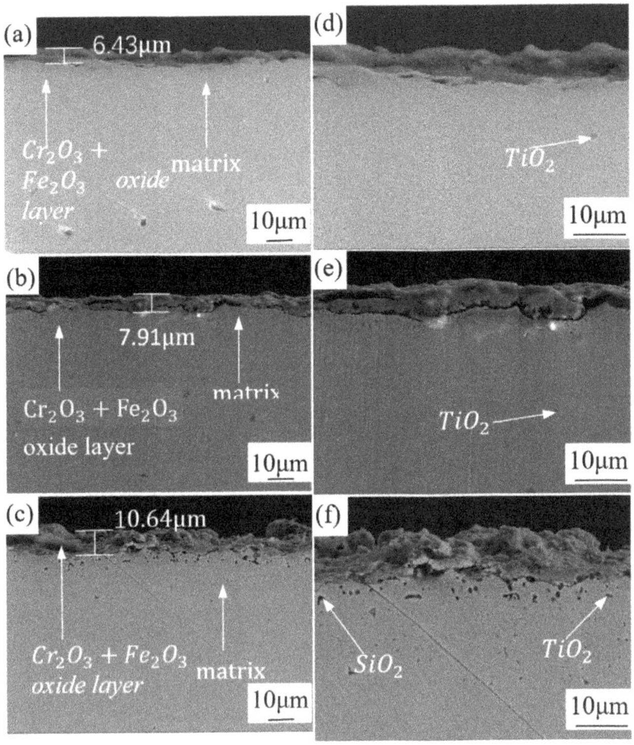

Figure 6. SEM morphology of 441 stainless steel at the interface between substrate and oxide layers at different temperatures (**a**,**d**) 700 °C, (**b**,**e**) 800 °C, (**c**,**f**) 900 °C.

Comparing the oxidation cross-sections of two types of stainless steel, the oxide layer thickness of 441 stainless steel is smaller than that of 439 stainless steel, and the integrity of the oxide layer is better, reflecting that 441 stainless steel has stronger high-temperature oxidation resistance than 439 stainless steel. In addition, comparing Figure 5f with Figure 6f, it can be found that SiO_2 particles are generated between the surface of the oxidation layer and metal matrix after oxidation at 900 °C. However, the size and density of the SiO_2 particles in 439 steel are greater than those of 441 steel. Research has shown that the SiO_2 layer can block the diffusion of oxygen to the matrix metal, so the more severe inward oxidation of 439 results in more SiO_2 particles generated at high temperatures than in 441 stainless steel [12].

Oxygen can penetrate into the stainless steel matrix, and Ti and Si that were originally dissolved near the surface of the stainless steel can combine with oxygen to generate TiO_2 and SiO_2. This makes the concentration of Ti and Si solution decrease. Under increased diffusion ability of the crystal interface at high temperature, according to Fick's first law, Ti and Si inside the matrix begin to diffuse towards the oxide film, and a large number of TiO_2 and SiO_2 particles are generated near the junction of the matrix and the oxide film. Finally, voids are formed around the particles due to the Kirkendall effect.

3.3. Analysis of Composition Distribution and Phase Structure of the Oxidized Surface at Different Temperatures

In order to study phase changes during high-temperature oxidation, EDS was used to scan the oxidized surface to analyze elemental distribution (Figures 7 and 8), and XRD phase composition analysis was performed on the oxidized surface (Figure 9).

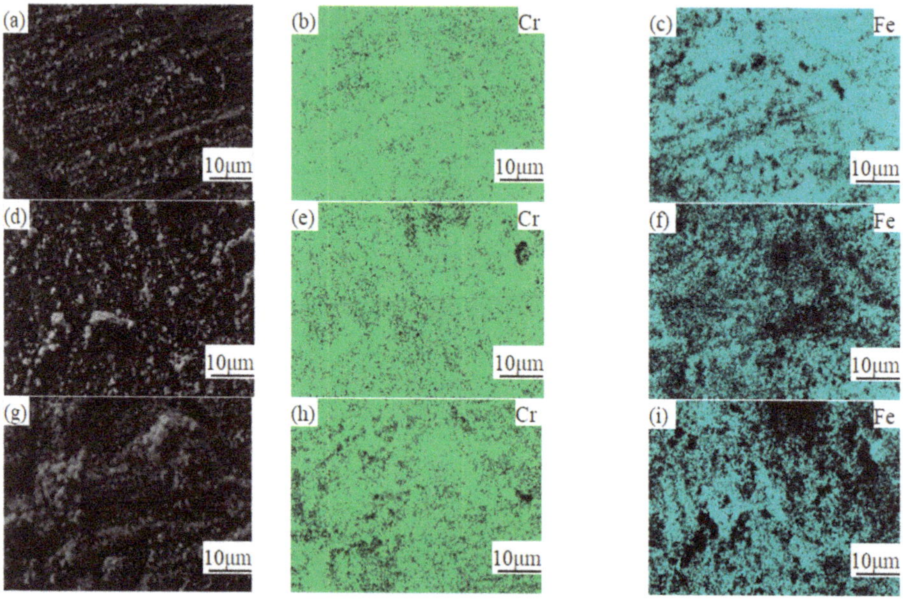

Figure 7. Element distribution on 439 high-temperature oxidation surface at different temperatures (EDS surface scanning) (**a**) 700 °C-morphology, (**b**) 700 °C-Cr, (**c**) 700 °C-Fe, (**d**) 800 °C-morphology, (**e**) 800 °C-Cr, (**f**) 800 °C-Fe, (**g**) 900 °C-morphology, (**h**) 900 °C-Cr, (**i**) 900 °C-Fe.

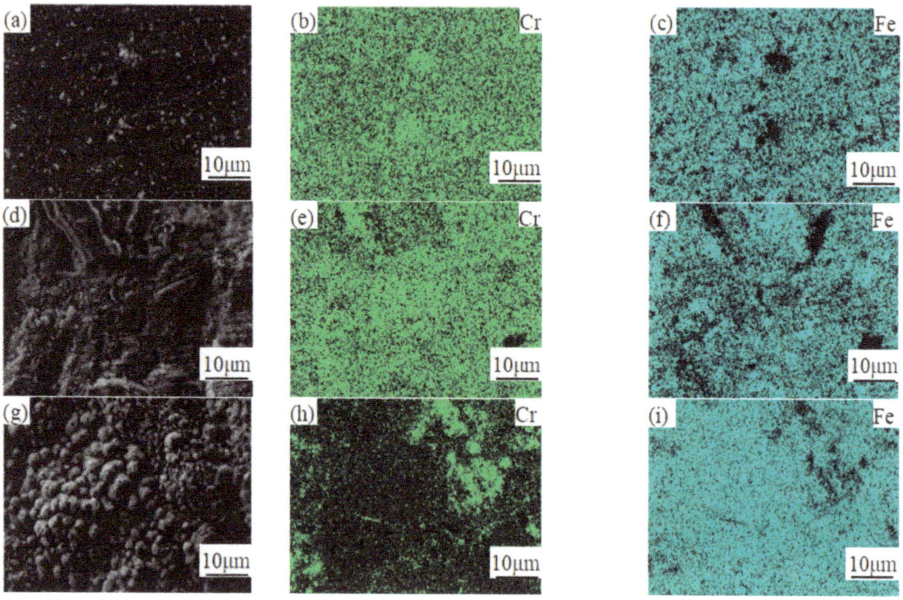

Figure 8. Element distribution on 441 high-temperature oxidation surface at different temperatures (EDS surface scanning) (**a**) 700 °C-morphology, (**b**) 700 °C-Cr, (**c**) 700 °C-Fe, (**d**) 800 °C-morphology, (**e**) 800 °C-Cr, (**f**) 800 °C-Fe, (**g**) 900 °C-morphology, (**h**) 900 °C-Cr, (**i**) 900 °C-Fe.

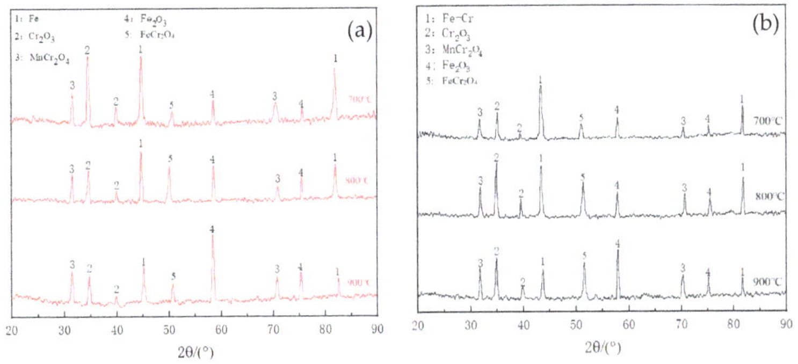

Figure 9. XRD spectra of oxidation products on the surface of two types of stainless steel after high-temperature oxidation (**a**) 439, (**b**) 441.

Figure 7 shows the distribution of oxidation surface composition of 439 stainless steel. From Figures 7b,e and 9, it can be seen that the oxidation layer of 439 stainless steel in the early stage of oxidation is mainly composed of Cr_2O_3. When the temperature increases from 700 °C to 800 °C, the distribution of the Cr element does not change significantly. Based on Figure 9a, it can be seen that the surface of the oxide layer is mainly composed of Cr_2O_3 and Fe_2O_3, but the specific morphology of the oxide layer is different. But when the oxidation temperature rises to 900 °C (Figure 7h), surface Cr rapidly decreases, and it is analyzed together with the experimental data in Figure 9a, only Fe_2O_3 remains on the surface oxide layer. The specific composition of the oxide layer is Fe_2O_3 and Cr_2O_3, mainly controlled by the diffusion of Cr element towards the surface layer of the matrix [13]. When 439 stainless steel is oxidized at a temperature of 900 °C, the Cr content on the oxidized surface significantly decreases, and the Cr element on the surface escapes from the oxide layer through the following reactions [14]:

$$Cr_2O_3(s) + \frac{3}{2}O_2(g) = 2CrO_3(g) \quad (1)$$

The generated CrO_3 is prone to volatilization at high temperatures, so the Cr_2O_3 and Fe_2O_3 mixed solid oxide layer that should have been generated only leaves a shell of Fe_2O_3. Solid Cr_2O_3 in the mixture is transformed into a gaseous CrO_3, which evaporates from pores in the shell, as shown in Figure 2e. At the same time, the hollow shell structure is more prone to cracking under internal stress.

By analyzing the specific composition of the oxide layer based on the XRD spectrum in Figure 9, it can be seen that when the oxidation temperature is 700 °C, the oxidation products of 439 stainless steel, except for the mixed oxide layer composed of Cr_2O_3 and Fe_2O_3, contain $MnCr_2O_4$ spinel particles in the oxide layer. The oxide film generated at lower temperatures is relatively thin, and Fe and Cr in the matrix are also detected. The two exist in a solid solution manner in the collective, making it difficult to distinguish them. After the temperature rises to 800 °C, an increase in surface oxide film thickness can be observed at the macro level, while an increase in Cr_2O_3 peak intensity at the micro level indicates that the oxide film generated by 439 stainless steel is thicker. In addition to the increase in thickness of the oxide layer, some areas exhibit a Fe_2O_3 shell structure during sintering, resulting in an increase in the overall Fe_2O_3 peak strength. Moreover, during the oxidation cycle, the oxide layer is prone to cracking and peeling due to thermal stress, and the protective ability of the matrix is lost in these areas. When the Cr element is sufficient, the cracked and peeled parts regenerate an oxide film, resulting in an overall increase in content. In addition to Cr_2O_3, it can also lead to oxidation of the matrix to form Fe_2O_3 or other oxides. As the oxidation temperature rose to 900 °C, the Cr_2O_3 component in the oxide film decreased significantly, while the iron oxide Fe_2O_3 increased significantly. Stainless steel can protect the substrate by adhering Cr_2O_3 on its surface in non-high-temperature working environments. However, under high temperature conditions such as 900 °C in this experiment, Cr_2O_3 will react with oxygen, resulting in CrO_3, which is a volatile oxide at high temperatures [15]. In this experiment, due to the transformation of 439 stainless steel during oxidation at 900 °C, the CrO_3 volatilization generated by the oxide layer significantly reduced the surface Cr content (Figure 7h). This allows the external environment to come into contact with the substrate through the generated pores, providing a way for further inward oxidation [16].

The EDS spectra of 441 stainless steel show little difference from 439 stainless steel at 700 °C and 800 °C, but there is a significant difference at 900 °C. The surface has a higher Cr content and a lower Fe content. The reason is that 439 stainless steel experienced a significant loss of Cr on the reaction surface, resulting in iron element in the oxide layer increased covering the surface [17]. However, 441 stainless steel did not undergo this reaction, so there is still a mixed iron chromium oxide layer on the surface. According to the XRD phase analysis in Figure 9, it can be seen that there are fewer oxidation products on the surface of 441 stainless steel compared to 439 stainless steel at an oxidation temperature of 700 °C. In Figure 9a,b, the oxidation does not clearly reflect the stronger high-temperature oxidation resistance of 441. However, in Figure 9c, the content of Cr_2O_3 in 441 is actually higher than that in 439, indicating a significant loss of Cr on the surface of 439 and the loss of the protective ability of the oxide layer.

According to the XRD phase analysis of two types of stainless steel cross-sections in Figure 10, the main reason for the different morphology of the oxide layer is the effect of niobium element. Niobium can refine ferrite grains [18]. From Figure 10a at 700 °C, Figure 10b at 800 °C, and Figure 10c at 900 °C, it can be observed that the content of Cr_2O_3 decreases, while the content of Fe_2O_3 increases, and the change in 439 stainless steel is more significant than that in 441 stainless steel. To enable the oxide layer to protect the substrate, it is necessary to ensure the stable growth of Cr_2O_3 during the oxidation process. It is necessary to ensure that the chromium flux diffused from the substrate to the oxide interface is greater than the chromium flux diffused from the oxide film to the outside. Otherwise, it is easy to cause chromium depletion. The Cr_2O_3 reaction in the Cr_2O_3, Fe_2O_3 and TiO_2 mixed oxide layer is depleted, resulting in voids like Figure 7c. At this point, the chromium content supplied by the matrix is insufficient, so Cr_2O_3 cannot be formed again on the matrix without oxide layer protection in the cavity, and the matrix will directly participate in oxidation [19].

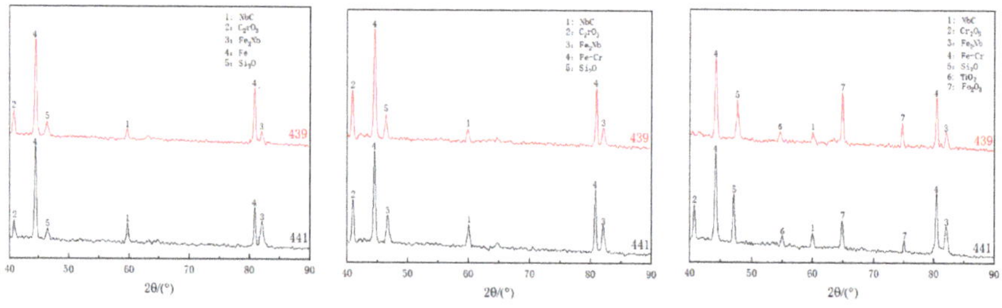

(**a**) (**b**) (**c**)

Figure 10. XRD phase analysis of high-temperature oxidation cross-sections of ferritic stainless steel at different temperatures (**a**) 700 °C, (**b**) 800 °C, (**c**) 900 °C.

In order to analyze the impact of precipitates on the resistance to high-temperature oxidation performance, typical morphologies of precipitates in two samples, Figures 11 and 12, were observed using TEM. According to the morphology and energy spectrum analysis of the precipitates, it can be determined that the precipitates in the 439 sample with low niobium content are mainly composed of (Nb, Ti) C composite precipitates within the crystal. The size is relatively large; the diameter of precipitates ranges from 2 to 3 μm;

however, no large particles were found in the 441 sample with a higher niobium content. Compared to the precipitates in 439 stainless steel, the precipitates on the grain boundaries of 441 stainless steel are finer; the diameter of precipitates ranges from 100 to 200 nm. As shown in Figure 12b–d, the composition of the precipitates includes Nb and Ti instead of C, which can be inferred as (Nb, Ti) N composite precipitation. Due to the low content of N and the difficulty of scanning N elements, this elemental energy spectrum appears [20].

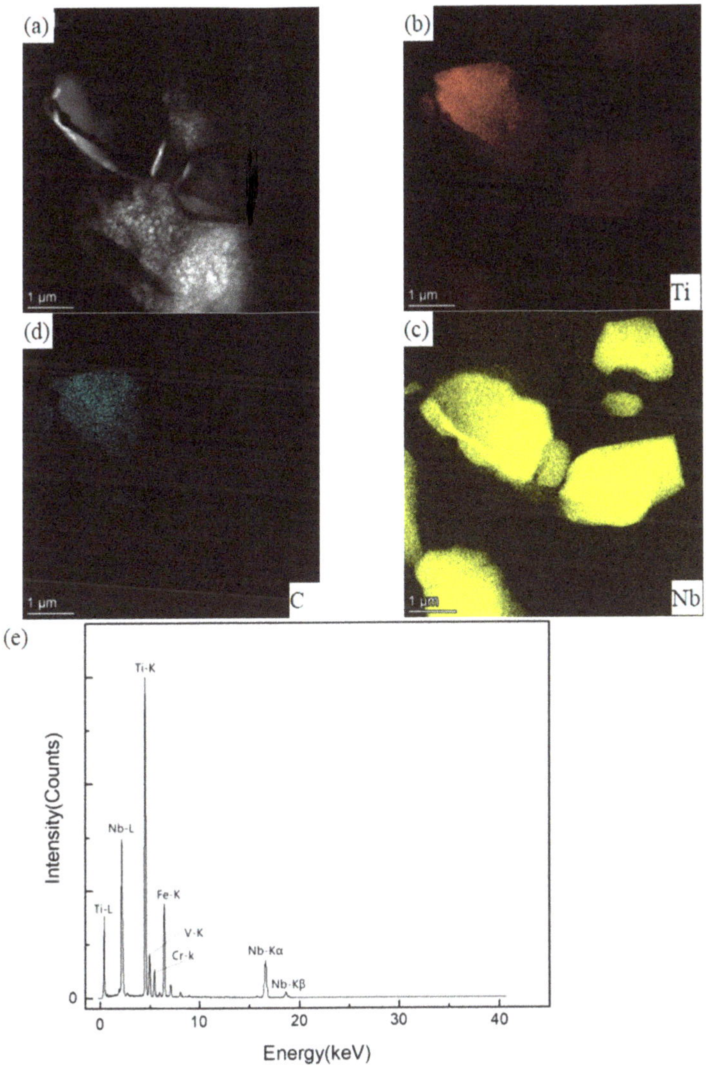

Figure 11. TEM morphology and elemental analysis of 439 stainless steel precipitates after oxidation at 900 °C (**a**) TEM morphology map; (**b**) Ti elemental energy spectrum; (**c**) energy spectrum of Nb element; (**d**) C element energy spectrum; (**e**) elemental spectrogram.

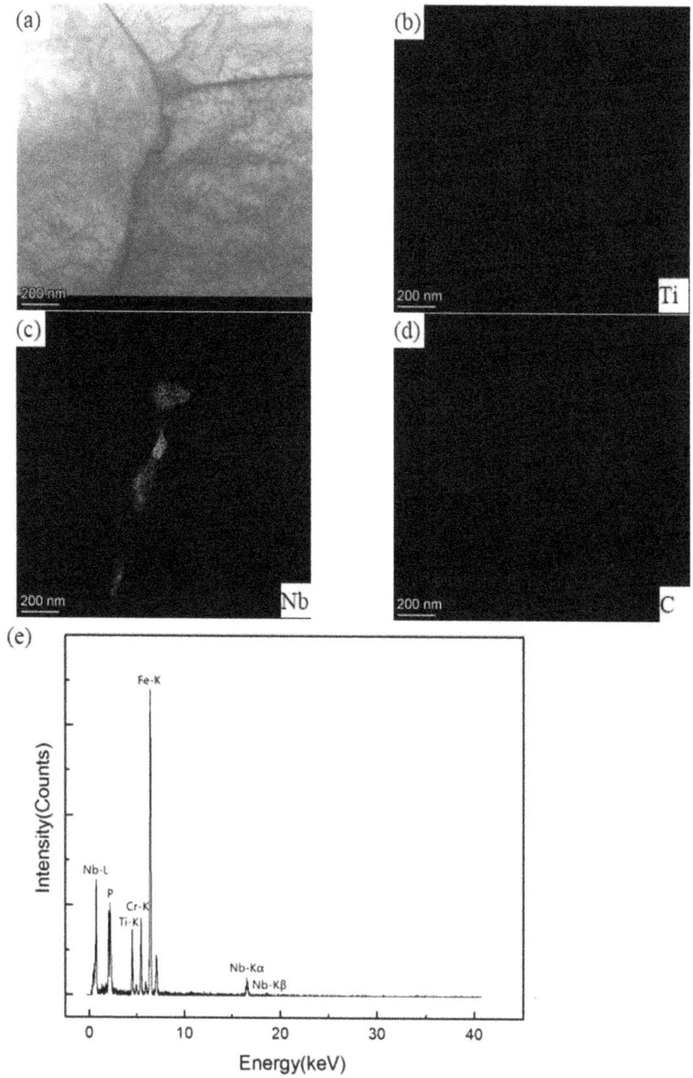

Figure 12. TEM morphology and elemental analysis of 441 stainless steel precipitates after oxidation at 900 °C (**a**) TEM morphology map; (**b**) Ti elemental energy spectrum; (**c**) energy spectrum of Nb element; (**d**) C element energy spectrum; (**e**) elemental spectrogram.

As shown in Table 1, 439 stainless steel has a high content of Ti and Al. Due to the higher precipitation temperature of TiN and AlN compared to NbN during melting, TiN and AlN precipitate before NbN and consume most of the N element. Therefore, Nb elements in 439 stainless steel are more likely to form NbC composite precipitates. In 441 stainless steel, the initial precipitation of TiN provides nucleation for NbN precipitation, resulting in (Nb, Ti) N composite precipitation.

4. Discussion

During the oxidation process, Gibbs free energy decreases with the increase of temperature, and the decrease in interfacial free energy between carbides and the matrix promotes the aggregation and growth of carbides [21]. On the other hand, the diffusion caused by the concentration difference between carbides and alloy elements in the matrix provides a dynamic driving force. Therefore, the 439 stainless steel before the high-temperature oxidation experiment did not exhibit 2–3 properties as shown in Figure 10a μm long (Nb, Ti) C composite precipitates. However, during high-temperature oxidation, the carbides undergo coarsening as the temperature increases.

Before high-temperature oxidation, the average particle size of precipitates is balanced locally between particles and crystals at the boundary of precipitates. When high-temperature oxidation occurs, Gibbs free energy decreases with the increase of temperature, and large-sized precipitates have smaller Gibbs energy addition than average sized precipitates. In order to maintain the overall composition of the precipitates, the size of the precipitates will correspondingly increase [22].

However, no large particle precipitates were found in 441, and the precipitated fine NbN played a fixed role at the grain boundary after precipitation, thus preventing further grain growth and reducing the average grain size. Grain refinement improves the oxidation resistance of steel, reduces its oxidation weight gain, and delays the occurrence of unstable oxidation. The higher content of niobium in 441 stainless steel is more likely to combine with C and N, which to some extent prevents the combination of Cr element with N and C to generate CrN and Cr_3C_2 [23]. During high-temperature oxidation of stainless steel, it is easy to generate CrN and Cr_3C_2 deposits at grain boundaries, and the areas immediately adjacent to CrN will lose chromium elements and form chromium-deficient areas. If a chromium-deficient area appears, it will increase the activity of chromium elements near the area, leading to the loss of Cr element during high-temperature oxidation [24].

Firstly, during the initial process of oxidation, the stainless steel oxide film will begin to form pores or microcracks, leading to the growth of some oxide grains in the oxide film, resulting in the appearance of oxidized surface particles as shown in Figures 2a and 3a. When oxidized at high temperature, O element diffuses inward, and the solid-soluble Cr in the matrix will preferentially combine with O to form Cr_2O_3 [25]. When the content of Cr in the matrix is insufficient, O will combine with F to form Fe_2O_3, causing failure. By adding Nb to fix Cr, Nb will preferentially combine with N and C in steel to form Nb (N, C) and other precipitates, thus avoiding the decrease of the concentration of solid-soluble Cr in the matrix. As the temperature increases, the Cr_2O_3 layer in the oxidation layer of 439 stainless steel continuously evaporates, resulting in a large number of defects such as oxidation layer pores and cracks on the surface, leading to continuous inward oxidation. The speed of supplying Cr from the substrate to the oxide skin through diffusion is not fast enough to maintain a high Cr concentration in the oxide layer, and Fe is easily diffused outward, forming Fe-rich nodular oxides [26].

As mentioned above, no Nb-containing intergranular precipitates were found in the 439 steel as shown in Figure 12a, resulting in chromium-deficient areas. At high temperatures, the Cr element further reacts to generate volatile CrO_3, resulting in voids in the previous solid oxide layer caused by volatilization of CrO_3. Through the voids, the matrix can continue to come into contact with the external environment. Therefore, intense oxidation reactions will continue to occur near the chromium-rich area on the substrate, resulting in a thicker oxide layer on the side near the substrate in the cavity shown in Figure 4f.

Overall, the (Nb, Ti) C precipitates generated by 439 stainless steel will continue to grow during high-temperature oxidation, which is detrimental to the high-temperature performance of stainless steel [27]. The (Nb, Ti) N precipitates generated between grains in 441 stainless steel can play a role in pinning crystals to prevent grain coarsening during high-temperature oxidation and solidify Cr [28], so the high-temperature oxidation resistance of 441 stainless steel is superior to that of 439 stainless steel.

5. Conclusions

(1) As the oxidation temperature increases, the oxide layer of both types of stainless steel will increase. The oxide layer thickness of 441 stainless steel with higher niobium content is less than 439 stainless steel at the same temperature. The oxide layer of 441 stainless steel is intact at any temperature, while 439 stainless steel exhibits cracking and voids in the oxide layer at high temperatures.

(2) The phase structure composition and oxidation morphology of the oxide layer vary at different temperatures. When the sample is oxidized at 700 °C, the oxidation layer of both types of ferritic stainless steel are Cr_2O_3 and Fe_2O_3. The oxidation layer of 439 stainless steel is dense and uniform, while 441 stainless steel is a discrete oxide particle. When the sample is oxidized at 800 °C, the oxidation products on the surface of 439 stainless steel are sintered together, and needle-like Fe_2O_3 structures appear on the oxide film, while the oxidation particles on the surface of 441 stainless steel grow and aggregate into a uniform oxide layer. When the sample is oxidized at 900 °C, the oxidation layer exhibits internal and external layering, with an outer layer of Fe_2O_3 shell structure and an inner layer of iron-containing oxides such as Fe_2O_3, Fe_3O_4, and FeO. However, 441 stainless steel has a needle-like Fe_2O_3 oxidation tumor on its surface but no cracking occurs, indicating a relatively complete oxidation layer.

(3) The reason why 439 steel has poor high-temperature oxidation resistance is that (Nb, Ti) C precipitates coarsen at high temperatures, which has a destructive effect on high-temperature resistance. The reason why 441 steel has good high-temperature oxidation resistance is due to the precipitation of NbN and other niobium-containing precipitates at grain boundaries, which has the effect of refining grains and hindering the escape of Cr elements.

Author Contributions: Conceptualization, E.Z., D.X., P.C. and Q.H.; methodology, D.X.; validation, X.W., Z.L. and K.J.; investigation, D.X. and Z.L.; writing—original draft preparation, E.Z., D.X. and Q.H.; writing—review and editing, D.X.; project administration, D.X. and Q.H.; funding acquisition, P.C. All authors have read and agreed to the published version of the manuscript.

Funding: Shandong Provincial Natural Science Foundation, ZR2021ME247. Major R&D Plan Project in Shandong Province, 2023CXPT081.

Data Availability Statement: The data presented in this study are available on request from the corresponding author. As this paper is based on sponsored Funding, experimental data are used for the research and development of new ferritic stainless steel in Funding, so experimental data are not publicly available.

Conflicts of Interest: Author Peidun Chen was employed by the company Shandong Taishan Iron and Steel Group Co. The remaining authors declare that the research was conducted in the absence of any commercial or financial relationships that could be construed as a potential conflict of interest.

References

1. Oku, M. Development of Heat-resistant Ferritic Stainless Steel for Automotive Exhaust Systems Made by Nippon Steel. *Wo* *Steel* **2011**, *5*, 10–16.
2. Di Schino, J.M.; Barteri, K.M. High temperature resistance of a high nitrogen and low nickel austenitic stainless steel. *J. Mat. S* *Lett.* **2003**, *22*, 691–693.
3. Dong, Z.; Chen, H.; Lang, Y. The effect of niobium and titanium on the high-temperature mechanical properties of ferritic stainle steel used in the hot end of automotive exhaust systems. *Steel* **2013**, *48*, 64–68.
4. Liu, T.; Chen, L.; Bi, H.; Che, X. Effect of Mo on High-Temperature Fatigue Behavior of 15CrNbTi Ferritic Stainless Steel. *Ac* *Metall. Sin. Engl. Lett.* **2014**, *27*, 452–456. [CrossRef]
5. Meng, Q.; Li, D.; Yang, J.; Liu, T. High temperature oxidation behavior of 310S heat-resistant steel. *Mater. Eng.* **2022**, *50*, 137–1
6. Zhang, M.; Han, Y.; Zu, G.; Sun, J.; Zhu, W.; Chen, H.; Ran, X. High-Temperature Oxidation Behavior of a Cu-Bearing 17 Ferritic Stainless Steel. *Scanning* **2020**, *2020*, 8847831. [CrossRef]
7. Badin, V.; Diamanti, E.; Forêt, P.; Darque-Caretti, E. Characterization of oxide scales formed on ferritic stainless steel 441 1100 °C under water vapor. *Oxid. Met.* **2014**, *82*, 347. [CrossRef]
8. Movahedi-Rad, A.; Pelaseyed, S.S.; Attarian, M.; Shokrallahzadeh, R. Oxidation behavior of AISI 321, AISI 316, and AISI 4 stainless steels: Kinetic, thermodynamic, and diffusion studies. *J. Mater. Res.* **2016**, *31*, 2088–2096. [CrossRef]

Jacob, Y.P.; Haanappel, V.; Stroosnijder, M.F. The effect of gas composition on the isothermal oxidation behaviour of PM chromium. *Corros. Sci.* **2002**, *44*, 2027–2039. [CrossRef]
Raj, K.A. On High-Temperature Materials: A Case on Creep and Oxidation of a Fully Austenitic Heat-Resistant Superalloy Stainless Steel Sheet. *J. Mater.* **2013**, *2013*, 124649.
Cheng, L. Study on High Temperature Oxidation Behavior of Ferritic Stainless Steel. Ph.D. Thesis, Shenyang University, Shenyang, China, 2021.
Duan, X.; Zhang, J.; Zhang, S.; Li, G.; Wei, Y. The effect of Nb and Ti on the high-temperature oxidation behavior of 444 ultra pure ferrite stainless steel. *Spec. Steel* **2021**, *42*, 7–11.
Seo, H.S.; Yun, D.W.; Kim, K.Y. Oxidation behavior of ferritic stainless steel containing Nb, Nb-Si and Nb-Ti for SOFC interconnect. *Int. J. Hydrogen Energy* **2012**, *38*, 2432–2442. [CrossRef]
Safikhani, A.; Esmailian, M.; Tinatiseresht, T.; Darban, G.B.D. High temperature cyclic oxidation behavior of ferritic stainless steel with addition of alloying elements Nb and Ti for use in SOFCs interconnect. *Int. J. Hydrogen Energy* **2016**, *41*, 6045–6052. [CrossRef]
Haitao, Y.; Hongyun, B.; Xin, L.; Zhou, X. Precipitation and mechanical properties of Nb-modified ferritic stainless steel during isothermal aging. *Mater. Charact.* **2009**, *60*, 204.
Peng, J.; Li, M.; Luo, S. Study on high-temperature oxidation behavior of B443NT ferritic stainless steel. *Baosteel Technol.* **2013**, 22–25+31.
Huang, H.; Yang, Y. Study on high-temperature oxidation behavior of niobium titanium double stabilized 430 stainless steel. *J. Mater. Res.* **2014**, *28*, 641–648.
Fujita, N.; Ohmura, K.; Kikuchi, M.; Suzuki, T.; Funaki, S.; Hiroshige, I. Effect of Nb on high-temperature properties for ferritic stainless steel. *Scr. Mater.* **1996**, *35*, 705–710. [CrossRef]
Cheng, X.W.; Jiang, Z.Y.; Monaghan, B.J.; Longbottom, R.J.; Wei, D.B.; Hee, A.C.; Jiang, L.Z. Degradation of ferritic stainless steels at 1200 °C in air. *Mater. Corros.* **2018**, *69*, 63–75. [CrossRef]
Teng, S.; Qian, Z. Study on the high-temperature oxidation resistance of ultra-pure ferrite stainless steel. *Gansu Metall.* **2022**, *44*, 53–55.
Qin, Z.; Taosha, L.; Wen, W.; Langlang, Z.; Yuefeng, J. High temperature oxidation kinetics and evolution of oxide film structure of nanocrystalline Ni-12Cr alloy at 800 °C. *Chin. J. Corros. Prot.* **2022**, *42*, 733–742.
Chen, X.; Xiaochun, W.; Na, M.; Qianliang, S. 3DAP Study on the C Segregation Behavior of High Carbon High Alloy Steel During Deep Cooling Treatment. *J. Met.* **2015**, *51*, 325–332.
Huang, Z.; Yang, Y. The effect of niobium titanium on the high-temperature precipitation behavior of the second phase in ferritic stainless steel. *Met. Heat Treat.* **2015**, *40*, 7–12.
Xin, Z.; Wei, D.; Xunzheng, H. The effect of niobium titanium on the second phase precipitation and mechanical properties of 430 ferrite stainless steel. *Baosteel Technol.* **2015**, 6–12.
Chen, Y.; Zhan, J.; Ni, Q.; Li, M. Oxidation Behavior of 441 Ferritic Stainless Steel in Simulated Automotive Exhaust Gas Environment. *Corros. Sci. Prot. Technol.* **2019**, *31*, 174–180.
Zhao, T.; Teng, L.; Jin, Y.; Zhang, J.; Yang, Y. Microstructure and mechanical properties of niobium containing ferritic stainless steel suitable for automotive exhaust manifolds. *Shanghai Met.* **2020**, *42*, 63–68.
Hou, Q.; Jing, D.; Liao, Z.; Wang, K.; Xi, B.; Huang, Z. Effect of niobium on oxidation weight gain of 65SiCrV6 spring steel. *Steel* **2022**, 1–15.
Duan, X.; Zhang, J.; Wang, B.; Li, G. Behavior of Trace Elements during High Temperature Oxidation of 444 Ferritic Stainless Steel. In Proceedings of the 13th China Steel Annual Conference-7. Advanced Steel Materials and Their Applications, Online, 23–24 November 2022; pp. 443–449.

claimer/Publisher's Note: The statements, opinions and data contained in all publications are solely those of the individual ıor(s) and contributor(s) and not of MDPI and/or the editor(s). MDPI and/or the editor(s) disclaim responsibility for any injury to ple or property resulting from any ideas, methods, instructions or products referred to in the content.

metals MDPI

Article

Characterization of Various Stainless Steels Containing Gadolinium as Thermal Neutron Absorbing and Shielding Materials

SeKwon Oh [1], Ji-Ho Ahn [1], Rockhoon Jung [2], Hyun-Jong Kim [1], Younghwan Chu [3], Dae Hyun Choi [4], Hyun Lee [5] and Hyun-Do Jung [5,6,*]

1. Research Institute of Advanced Manufacturing Technology, Korea Institute of Industrial Technology, Incheon 21999, Republic of Korea; sk0514@kitech.re.kr (S.O.); dkswlgh8@kitech.re.kr (J.-H.A.); hjkim23@kitech.re.kr (H.-J.K.)
2. Ship & Offshore Research Institute, Samsung Heavy Industries, Pangyo R&D Center, Seongnam-si 13486, Republic of Korea; rhoon.jung@samsung.com
3. Fospia Co., Ltd., Mado-ro, Mado-myeon, Hwaseong-si 18541, Republic of Korea; crystal@fospia.com
4. SHINWOO CNC KOREA Co., LTD, 87, Gongdan-ro, Goseong-eup, Goseong-gun 52948, Republic of Korea; ch240ch240@gmail.com
5. Department of Biomedical-Chemical Engineering, The Catholic University of Korea, Bucheon 14662, Republic of Korea; leeh0520@catholic.ac.kr
6. Department of Biotechnology, The Catholic University of Korea, Bucheon 14662, Republic of Korea

* Correspondence: hdjung@catholic.ac.kr; Tel.: +82-2-2164-4357

Abstract: Developing next-generation thermal-neutron-shielding and -absorbing materials for the safe storage and transportation of spent nuclear fuel is a topic of active research in academia. Gadolinium (Gd) boasts superior neutron absorption capacity compared with other nuclei. Consequently, it has garnered significant attention as a potential replacement for boron in spent nuclear fuel (SNF) applications. In this study, the austenitic stainless steels 304 and 316 and the duplex stainless steel 5A were reinforced with 2 wt.% of gadolinium to explore their thermal-neutron-absorbing efficiency. Their properties were then compared with commercial-grade borated stainless steel, as per ASTM standard A887. After the inclusion of Gd, both the yield strength and ultimate tensile strength of the alloys slightly increased, which was attributed to the evenly distributed Gd intermetallics within the matrix. However, the elongation rate was reduced. The Charpy impact absorption energy also decreased. The influence of the Gd intermetallic was further explored by examining the fractography. While the corrosion resistance of the alloy slightly diminished after the addition of Gd, its neutron absorption capacity demonstrated superior performance, especially when the Gd content was equivalent to that of boron. Although most properties of the experimental alloy deteriorated after the Gd addition, they still outperformed commercial borated stainless steel, suggesting that this alloy might be a promising candidate for SNF applications.

Keywords: gadolinium; stainless steel; neutron-absorbing material; mechanical properties; microstructure; corrosion resistance

1. Introduction

Spent nuclear fuel (SNF) generated by nuclear reactors has the inherent capability to undergo spontaneous fission, resulting in the emission of intense neutrons and gamma rays [1–4]. These emissions pose significant radiation risks to both individuals and the surrounding environment, necessitating careful and proper disposal. The safe management of SNF hinges on effective transportation and storage practices, with neutron-shielding and -absorbing materials serving as pivotal components [5,6].

Among these materials, boron steel and boron carbide (B_4C)/aluminum (Al) composite materials are notable examples, where they rely on the high thermal neutron absorption cross-section of boron-10 (^{10}B) [7]. These materials find application in the storage racks and transport casks used for SNF. To enhance the neutron-shielding and -absorbing effectiveness of boron steel, enriched ^{10}B is typically introduced into steel through alloying. This process, however, is intricate and costly. Moreover, the highest achievable boron (B) content in such steel remains limited to 2.25 wt% [8]. Unfortunately, boron steel with this B content cannot entirely shield thermal neutrons, and its mechanical properties significantly deteriorate beyond this concentration [4]. Consequently, it is deemed unsuitable for the secure transportation and storage of SNF. Furthermore, the borides within boron steel tend to form helium bubbles upon neutron absorption, leading to the degradation of shielding materials over extended storage periods, thereby posing safety risks [9].

On the other hand, B_4C/Al neutron-shielding materials can contain substantially higher B_4C content, up to 35 wt%. However, materials with such elevated B_4C content exhibit exceptional hardness, making them challenging to process and shape [10]. Additionally, B_4C/Al composites as thermal-neutron-absorbing and -shielding materials cannot function as standalone structural materials and require a boron steel cladding layer for structural support. This complex design results in containers with intricate structures and limited storage capacity, rendering them inconvenient for use, cleaning, and maintenance [11]. Evidently, there exists a trade-off between the content of neutron-absorbing elements and the properties of neutron-absorbing and -shielding materials.

Gadolinium ($^{155,157}Gd$) boasts the highest thermal neutron absorption cross-section value compared with other nuclei. Furthermore, when considering thermal neutron cross-section values, ^{157}Gd boasts an exceptional advantage, presenting an approximately 60-fold increase (240,000 b) compared with boron (^{10}B) (3810 b) [12]. At an equivalent content, Gd exhibits a thermal neutron absorption capacity more than four times greater than that of ^{10}B, and it comes at a significantly reduced cost, ranging from 1/12 to 1/5 of ^{10}B [13]. Furthermore, Gd effectively mitigates helium bubble formation during neutron absorption reactions, thus extending the service life of neutron-absorbing and -shielding materials [14]. Substituting Gd for B in the production of thermal-neutron-absorbing and -shielding materials presents a solution to the content–performance dilemma. This approach allows for achieving superior neutron-shielding rates and enhanced material properties with lower Gd content.

Stainless steels (SSs) and nickel-based alloys are well regarded for their formidable mechanical strength and robust corrosion resistance. As a result, there has been considerable exploration into the amalgamation of Gd with various materials, including nickel-based alloys [15,16], austenitic stainless steels [17–19], and duplex stainless steels (DSSs) [20,21]. These alloys demonstrate promise in terms of reactivity control and structural functionality, albeit at the expense of reduced toughness and ductility owing to the formation of Gd-rich precipitated intermetallics. In recent developments, there has been growing interest in novel alloys that incorporate both Gd isotopes, aiming to strike a balance between effective neutron absorption and structural integrity [22,23].

SS 304, SS 316, and ASTM A890 5A grade duplex stainless steel were selected as base materials. To introduce Gd, 2 wt. % was incorporated and the alloys were prepared using vacuum arc melting (VAM). For a detailed examination of the alloy microstructures and inclusions, a field-emission scanning electron microscope (FE-SEM) was employed. The phases and compositions of these alloys were determined using X-ray diffraction (XRD) and energy-dispersive X-ray spectroscopy (EDS). Furthermore, the mechanical properties were assessed using tensile and Charpy impact resistance testing, while the corrosion resistance was evaluated via electrochemical corrosion tests.

2. Experimental Procedures

2.1. Alloy Design and Production

The experimental alloys were fabricated with VAM using electrolytic iron (Fe), chromium (Cr), nickel (Ni), manganese (Mn), silicon (Si), ferro-molybdenum (Fe-60 wt%Mo), and ferrochromium nitride (Fe-60 wt%Cr-10 wt%N), and high-purity Gd (99.99%, Treibacher

Industrie AG, Hamburg, Germany). To prevent oxidation, the dissolution was conducted in a high-vacuum state of $\geq 1 \times 10^{-5}$ Torr and Argon (Ar) was used as the atmosphere gas of the chamber. In the case of the experimental alloy, remelting was performed five times by inverting each specimen up and down to prevent segregation and promote the uniform distribution of Gd.

In this study, SUS 304, SUS 316, and DSS 5A alloys (listed in ASTM standard A890) were selected as base materials and chemical compositions of the experimental alloys were analyzed using inductively coupled plasma optical emission spectroscopy (ICP-OES, iCAP-6300, Thermo Scientific, Amberg, Germany), and the results are shown in Table 1. The amount of Gd added was 2 wt.%, and the material property changes before and after the Gd addition were analyzed according to the amount of added Gd. Before the preparation of the experimental alloy, the solidification behavior was confirmed and the appropriate heat treatment temperature was determined using our previous studies.

Table 1. Chemical composition of the experimental alloys (wt.%).

Alloys	Fe	Cr	Ni	Mo	Mn	Si	N	Gd
304 (Aimed)	Bal.	18	10	-	1.3	0.8	0.1	2
304	Bal.	18.4	9.21	-	1.01	0.45	0.13	-
304-Gd	Bal.	18.1	9.44	-	1.13	0.62	0.12	1.76
316 (Aimed)	Bal.	18	12	2.4	1.3	0.8	0.1	2
316	Bal.	19.3	11.6	1.9	1.4	0.91	0.12	-
316-Gd	Bal.	18.3	10.9	2.12	1.19	0.77	0.14	2.13
5A (Aimed)	Bal.	25	7	4.5	1.3	0.8	0.3	2
5A	Bal.	25.4	7.17	4.23	1.16	0.65	0.25	-
5A-Gd	Bal.	25.8	7.63	4.34	1.22	0.78	0.28	1.96

The prepared experimental alloy was subjected to a solution heat treatment to remove the intermetallic compounds generated during solidification. In the case of 304 and 316 alloys, the solution heat treatment was performed at 1080 °C, and in the case of 5A, a solid solution heat treatment was performed at 1130 °C to create a ratio of α to γ phases equal to 45:55.

The evaluation of the neutron-absorbing capacity of the fabricated alloys in this study followed a previously established method [1]. In essence, we employed the Beer–Lambert Law to theoretically calculate the neutron transmission ratio, as expressed below [24]:

$$I/I_0 = e^{-N\sigma t} \quad (1)$$

Here, I_0 and I represent the neutron transmission intensity before and after passing through the absorber, respectively. N denotes the atom density, σ signifies the total cross-section, and t stands for the thickness of the neutron absorber. To determine the atom density (N) of each element (x), we used the following formula [1]:

$$N_x = \rho N_a n_x / M \quad (2)$$

where ρ represents the density of the compound, M stands for the molecular weight of the compound, N_a corresponds to Avogadro's number (6.022×10^{23} atoms/mol), and n_x indicates the number of atoms of element x within one compound. Table S1 shows the various parameters of Fe and Gd used to calculate the neutron transmission ratio. Since the neutron cross-section value of Gd is much higher compared with that of other elements composing steel, we performed the calculation using only Fe, which is the main component of steel, and Gd.

2.2. Microstructure Characterization

For the microstructural examinations, the experimental samples underwent a preparation process involving grinding with SiC sandpaper ranging from ~180 to 2000 grit, followed by polishing with diamond suspension slurries. The assessment of the experimental alloys, both with and without Gd, was conducted using the backscattered electron (BSE) mode of a Quanta 200 F field-emission scanning electron microscope (FE-SEM) from the FEI Company in the USA, both before and after the solution treatment. An energy-dispersive X-ray spectroscopy (EDS) attachment integrated with the FE-SEM was employed to analyze the elemental composition of the specimens. To characterize the alloy phases, X-ray diffraction (XRD) analysis was performed using a D8-discover instrument from Bruker in Germany with monochromatic Cu Kα radiation. The scanning spanned a diffraction angle range from 20° to 80°, operating at a rate of 1°/min. The α phase content was quantified employing a feritscope (FMP-30) manufactured by Fischer in Germany, boasting a remarkable α detection limit of 0.1%. Notably, each specimen was subjected to analysis across an average of 30 fields. Furthermore, the average sizes and areas of inclusions present in the alloys were assessed through the utilization of an inclusion analyzer (CTR 6000) produced by Leica in Berlin, Germany.

2.3. Mechanical Properties of the Experimental Alloys

To investigate the mechanical characteristics of the experimental alloys, including those with and without Gd, we conducted assessments of their tensile strength and Charpy impact resistance. Cylindrical specimens were subjected to elongation at a constant crosshead speed of 1 mm/min, following the guidelines outlined in the ASTM standard A370. This evaluation of tensile properties was carried out using an AGX 300KN testing machine manufactured by Shimadzu in Japan. Furthermore, Charpy impact tests, in accordance with the KS-0809 standard, were performed on V-notched specimens measuring $10 \times 10 \times 55$ mm^3. These tests were executed using a Charpy-type impact tester (SAID3) from Satec in the USA, all of which were conducted at a temperature of 25 °C. Subsequently, the fractured surfaces of the specimens following these tests were meticulously examined using an FE-SEM. To ensure the reliability of our findings, each mechanical test was conducted on more than five specimens for each alloy, allowing us to derive statistically significant outcomes. All quantitative data are presented as mean values accompanied by standard deviations. For the purpose of the statistical analysis, a one-way ANOVA was employed, with a significance threshold set at $p < 0.05$.

2.4. Electrochemical Corrosion Test

To assess the corrosion resistance of our experimental alloys immersed in artificial seawater, we employed an electrochemical potentiostat (VersaSTAT 3, AMETEK Scientific Instruments, Princeton, NJ, USA). The composition of the artificial seawater strictly adhered to the ASTM standard D1141–98, and its pH was adjusted to 8.2 for consistency.

For all electrochemical corrosion assessments, a conventional three-electrode cell setup was used, which featured a working electrode with a surface area of 1 cm^2, a platinum plate as the counter electrode, and an Ag/AgCl electrode functioning as the reference electrode. To ensure stability, each alloy sample was initially submerged in the artificial seawater for 30 min to achieve a steady open-circuit potential before the polarization scan commenced. Subsequently, we conducted a potential sweep ranging from −1000 to 1000 mV, operating at a scan rate of 0.1 mV/s. From the resulting potentiodynamic polarization curves, we calculated key parameters, such as the corrosion potential (Ecorr), corrosion current density (Icorr), and cathodic (βc) and anodic (βa) Tafel slopes. Additionally, the polarization resistance (Rp), which is essential for determining corrosion rates, was determined using the Stern–Geary equation [25]:

$$R_p = \frac{\beta_c \times \beta_a}{2.3 I_{corr}(\beta_c + \beta_a)} \tag{3}$$

3. Results and Discussion

3.1. Microstructural Characterization

All alloying elements, including Gd, were close to their target compositions. The 304, 304-Gd, 316, and 316-Gd alloys were heat treated at 1080 °C and the 5A and 5A-Gd alloys were treated at 1130 °C to remove intermetallic precipitates, such as σ phase compounds on the α, and obtain the ideal phase fraction. We tried to qualitatively analyze the phases of the experimental alloys using XRD analysis, and the results measured after the solution heat treatment are shown in Figure 1. In the case of the experimental alloys to which Gd was not added, no specific peak was observed and typical peaks of austenitic and DSS were observed. In the case of the 5A experimental alloy, the sigma (σ) phase and chi (χ) peaks, which are harmful intermetallic compounds, did not appear. In the cases of 304-Gd and 316-Gd, Gd-related peaks did not appear.

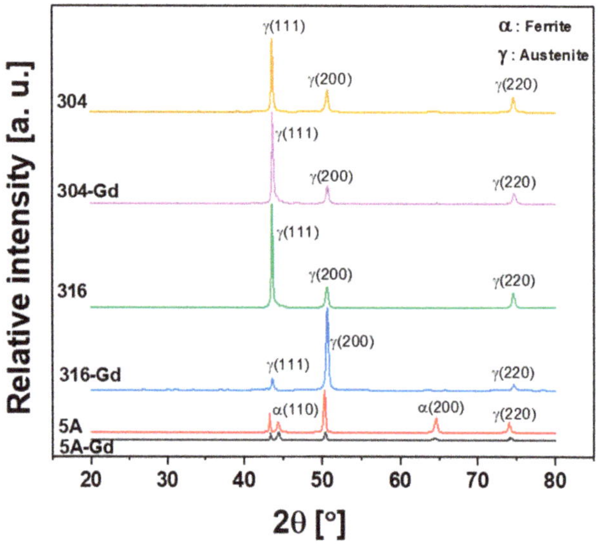

Figure 1. X-ray diffraction patterns of the experimental alloys after solution heat treatment.

Figure 2 shows SEM-BSE images after the solution heat treatment of 304-Gd, 316-Gd, and 5A-Gd. No microcracks or defects were found in any of the experimental alloys. The Gd intermetallic compounds showed a randomly distributed appearance throughout the specimen rather than a fixed appearance in a characteristic region or location. Also, no segregation or clusters were observed; there were no coarse intermetallic compounds. Because the experimental alloys were fabricated in a high vacuum, near 1×10^{-5} Torr, there were almost no oxides or inclusions. The composition of Gd intermetallics deposited in the experimental alloy was analyzed using SEM-EDS, and the results are shown in Figure 3. Most Gd intermetallics had a composition of Fe–Cr–Ni–Gd or Fe–Cr–Gd–N.

3.2. Mechanical Properties

The influence of Gd addition on the mechanical properties of the experimental alloys was examined by conducting tensile and Charpy impact tests. Figure 4 and Table 2 show the tensile test results. 5A showed higher tensile strength than 304 and 316 because it had a structure in which the α and γ phases were properly mixed. The ultimate tensile strength of 5A was 734 MPa, which was higher than that of the 304 and 316 alloys. The ultimate tensile strengths of 304-Gd, 316-Gd, and 5A-Gd were slightly increased compared with those of the alloys without Gd. The ultimate tensile strength of 304-Gd showed an improvement of approximately 7%, from 534 to 572.6 MPa, with Gd addition. The observed improvement was by approximately 11%, from 455 to 503 MPa, in the 316-Gd. The ultimate tensile strength of 5A-Gd was 776 MPa, which was an improvement of approximately 6% with the addition of Gd.

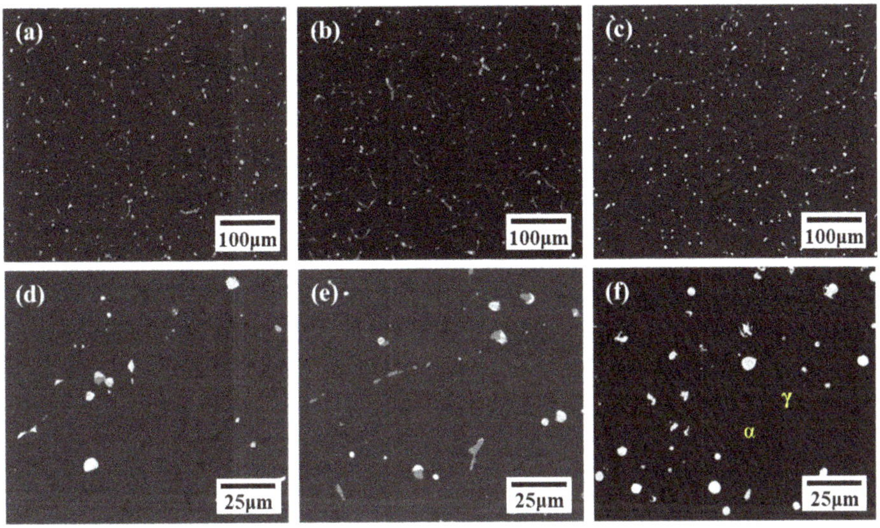

Figure 2. Representative backscattered electron (BSE) images of (**a**,**d**) 304-Gd, (**b**,**e**) 316-Gd, and (**c**,**f**) 5A-Gd. Upper images are low magnification and the lower images are high magnification.

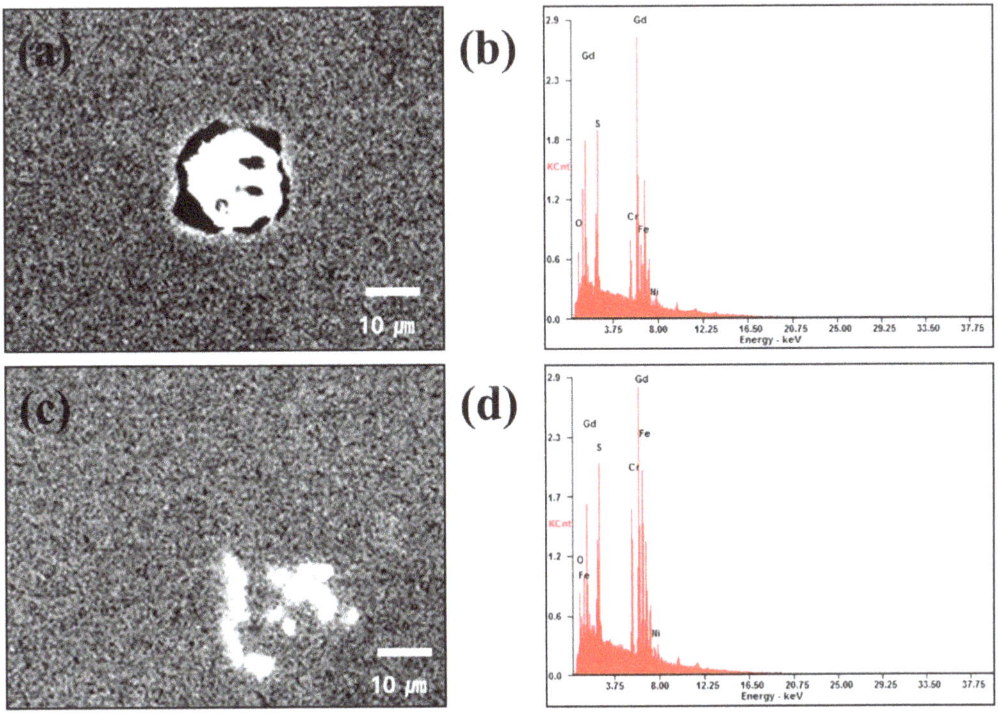

Figure 3. Representative (**a**,**c**) SEM images and (**b**,**d**) their EDS results of Gd-based intermetallics.

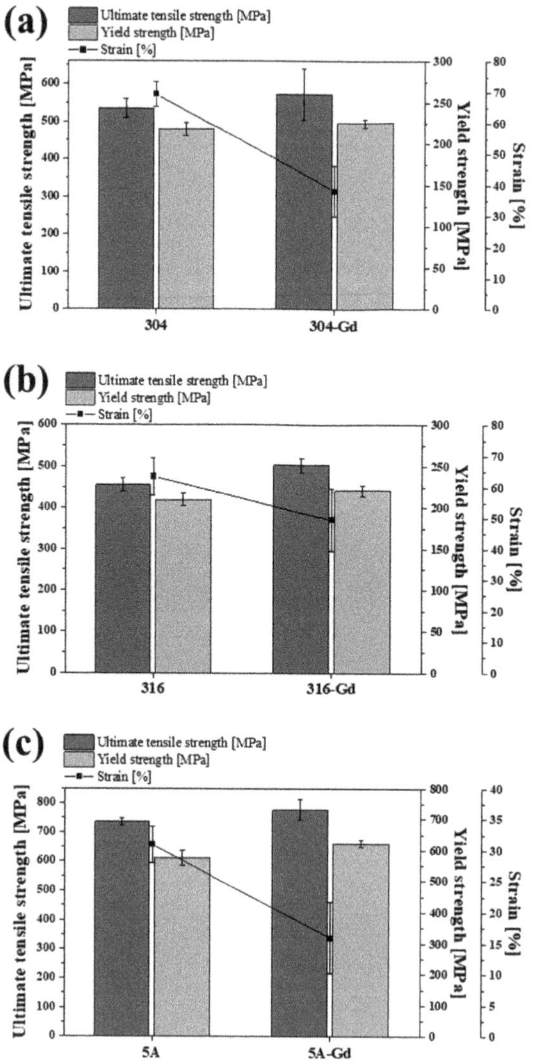

Figure 4. Changes in tensile properties of the experimental alloys with the addition of Gd: (**a**) 304 and 304-Gd, (**b**) 316 and 316-Gd, and (**c**) 5A and 5A-Gd.

The yield strength also slightly increased after Gd was added to the steels. In the case of 304, the yield strength was improved by approximately 4%, from 217 to 225 MPa. In the cases of 316 and 5A, yield strength improved by approximately 5%, from 209 to 220 MPa, and by approximately 8%, from 576 to 622 MPa, respectively. The improvements in ultimate tensile strength and yield strength are thought to be the result of the uniform distribution of Gd intermetallics generated by Gd addition to the matrix.

Conversely, the elongation of the Gd-added steels decreased significantly. In the case of 304, an approximately 44% decrease, from 69% to 38%, was observed after the addition of Gd. In the cases of 316 and 5A, the elongation decreased by approximately 22%, from 63% to 49%, and 48%, from 30% to 15.9%, respectively. In general, the intermetallics improved the strength of the alloy but reduced the toughness; therefore, the elongation decreased because of the presence of the Gd intermetallics.

Table 2. Tensile properties of the experimental alloys and borated stainless steel.

	UTS (MPa)	YS (MPa)	E (%)
304	534.2 ± 24.7	217.5 ± 7.4	69.3 ± 1.7
316	455.4 ± 16.3	210.0 ± 8.6	63.4 ± 2.7
5A	734.7 ± 10.1	576.1 ± 10.3	30.9 ± 2.6
304-Gd	572.6 ± 67.3	225.4 ± 6.8	38.3 ± 3.3
316-Gd	504.0 ± 15.6	220.9 ± 5.1	49.4 ± 4.6
5A-Gd	776.5 ± 24.3	622.4 ± 4.9	15.9 ± 5.1
304B6 (1.6 wt.% B)	515	205	9–20
304B7 (2 wt.% B)	515	205	6–17

Borated stainless steel, which is a representative thermal neutron absorber listed in ASTM standard A887, is made by adding B content ranging from 0.2 wt.% to a maximum of 2.25 wt.% to 304 (austenitic stainless steel). ASTM standard A887 has minimum mechanical property requirements according to the B content for use as a thermal neutron absorber. The results of the experimental alloys (304-Gd, 316-Gd, and 5A-Gd) were compared with the minimum mechanical property requirements of the ASTM standard [26–28]. Table 2 shows the tensile test results of the experimental alloys and the mechanical property requirements for 304B6 and 304B7 alloys with B contents of 1.6 and 2 wt.%, respectively. Generally, as the B content in the alloy increased, grain boundary segregation occurred and B intermetallics were nonuniformly dispersed. Therefore, as the amount of added B increased, the elongation decreased significantly. In the case of 304B7, in which the amount of B added was similar to the Gd amount of the experimental alloys, the required values of the ultimate tensile strength, yield strength, and elongation were 515 MPa, 205 MPa, and ~6–17%, respectively. The 304-Gd showed relatively high values for all three properties, with 316-Gd and 5A-Gd also exceeding the standard.

Figure 5 and Table 3 show the Charpy impact test results of the experimental alloys with solution heat treatment. 304, 316, and 5A showed impact energies of 125, 145, and 144 J, respectively, before Gd was added. In the case of 304-Gd (after adding Gd), the impact energy was reduced by approximately 49% to 64 J. The impact energy of 316-Gd decreased by approximately 56% to 62 J and 5A-Gd also showed a value of 77 J, which was approximately 46% lower than that before the Gd addition. The ASTM standard requires an impact energy of 22 J for 304B6, with 1.6 wt.% B added, and 14 J for 304B7, with 2 wt.% B added, as minimum values [28,29]. In the case of 304-Gd, which is an experimental alloy with a composition most similar to 304B7, the impact energy was 64 J, which is approximately 2.9 times higher than that of 304B7. Our findings align closely with prior research indicating that tensile properties, including ultimate tensile strength and strain at the point of fracture, exhibit a decline with increasing Gd content in the alloys. Qi et al. proposed that the proliferation of Gd-based intermetallic compounds significantly increases the interface between the second phase and the matrix [8]. This, in turn, results in a diminished bonding strength at this interface, rendering the material susceptible to crack initiation and propagation under tensile loads, ultimately leading to material fracture. Conversely, Wan et al. confirmed that an excessive accumulation of Gd-based intermetallics can trigger cracking during tensile testing [30]. This underscores the trade-off between augmenting neutron absorption capacity through increased Gd-based intermetallic content and impeding matrix connectivity, which, in turn, reduces material ductility.

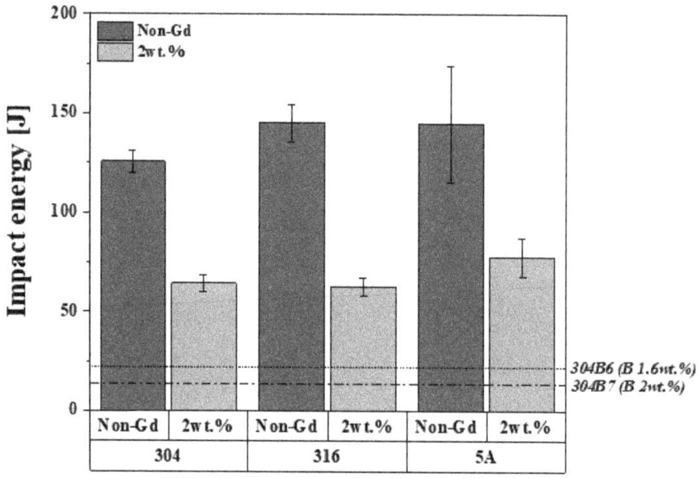

Figure 5. Changes in Charpy impact energy of the experimental alloys with the addition of Gd.

Table 3. Charpy impact energy of the experimental alloys and borated stainless steel.

Alloys	Impact Energy (J)
304	125.6 ± 5.1
316	145.4 ± 9.3
5A	144.8 ± 26.1
304-Gd	64.3 ± 3.8
316-Gd	62.8 ± 4.9
5A-Gd	77.7 ± 9.7
304B6 (1.6 wt.% B)	22
304B7 (2 wt.% B)	14

SEM fractographs of the experimental alloys after Charpy impact tests are shown in Figures 6–8. Figure 6 is the result of a low-magnification observation of the fracture surface. In the cases of Figure 6a–c, before the addition of Gd, ductile failure was observed. Conversely, ductile failure was observed after Gd was added, but cleavage failure was observed in some areas. Figure 7 shows images of the fracture surface observed at high magnification. In the cases of Figure 7a–c, before adding Gd, the sizes of the dimples were generally uniform, and the insides of the dimples were clean. Even after Gd was added, some brittle fractures were observed, but the dimples were evenly distributed overall, indicating a ductile fracture. However, Gd intermetallics were present inside the dimple and showed a broken shape. In general, precipitates or inclusions cause stress concentration and act as the main cause of lower shock energy absorption. In previous research, adding Gd in a small amount improved the impact properties by changing the inclusion morphology. However, in the experimental alloys in this study, stress was concentrated on the Gd intermetallics, and thus, it is thought that the dimple did not grow large and deep. The Gd intermetallics at the center of the dimple were analyzed using SEM-EDS, and the results are shown in Figure 8. The size and fracture patterns of the Gd intermetallics varied, with most of the chemical compositions being Fe–Cr–Ni–Gd-based intermetallics.

Figure 6. Low-magnification SEM images of the fracture surface of the experimental alloys: (**a**) 304, (**b**) 316, (**c**) 5A, (**d**) 304-Gd, (**e**) 316-Gd, and (**f**) 5A-Gd.

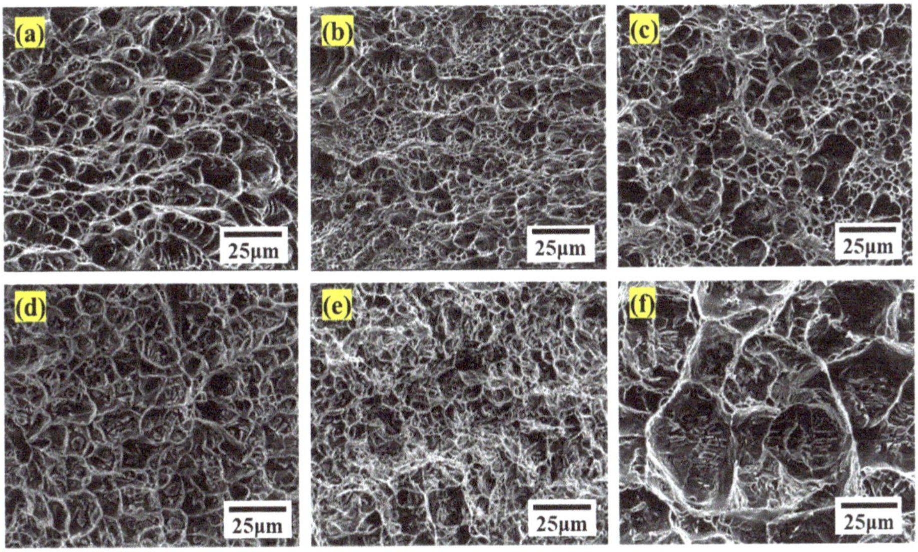

Figure 7. High-magnification SEM images of the fracture surface of the experimental alloys: (**a**) 304, (**b**) 316, (**c**) 5A, (**d**) 304-Gd, (**e**) 316-Gd, and (**f**) 5A-Gd.

3.3. Corrosion Resistances of the Alloys

Figure 9 and Table 4 show potentiodynamic polarization test results for the experimental alloys in an artificial seawater solution. After Gd was added, the polarization curves of all alloys showed a decrease in corrosion potential (E_{corr}) and an increase in corrosion current density (I_{corr}) compared with those of the alloys without the Gd addition. In general, the corrosion resistance increased as the E_{corr} value increased and the I_{corr} value decreased. In the case of the experimental alloy, the corrosion resistance decreased after adding Gd. However, as can be seen from the results in Table 4, the properties slightly decreased compared with those before the addition of Gd. In the case of 304, the R_p values were 107.17 and 33.206 before and after the Gd addition, respectively. In the cases of 316 and 5A, the R_p values were 133.12 and 35.391 and 119.69 and 17.02 before and after the Gd addition, respectively. The decreases in Rp value after the Gd addition were 69.01%, 74.41%, and 85.78% for 304, 316, and 5A, respectively. It is noteworthy that our results are similar to previous studies in that the corrosion properties of the Gd-incorporated alloys decreased as the contents of Gd increased [8,17,31]. This was attributed to the higher reactivity of Gd compared with Fe, Cr, and Ni. In the primary cell, Gd tended to dissolve at the negative electrode, leading to preferential corrosion of Gd-based intermetallic compounds. Additionally, the tendency of these Gd-based compounds to corrode more readily stemmed from Gd's heightened electrochemical activity relative to other matrix elements and its limited capacity to create protective passive films.

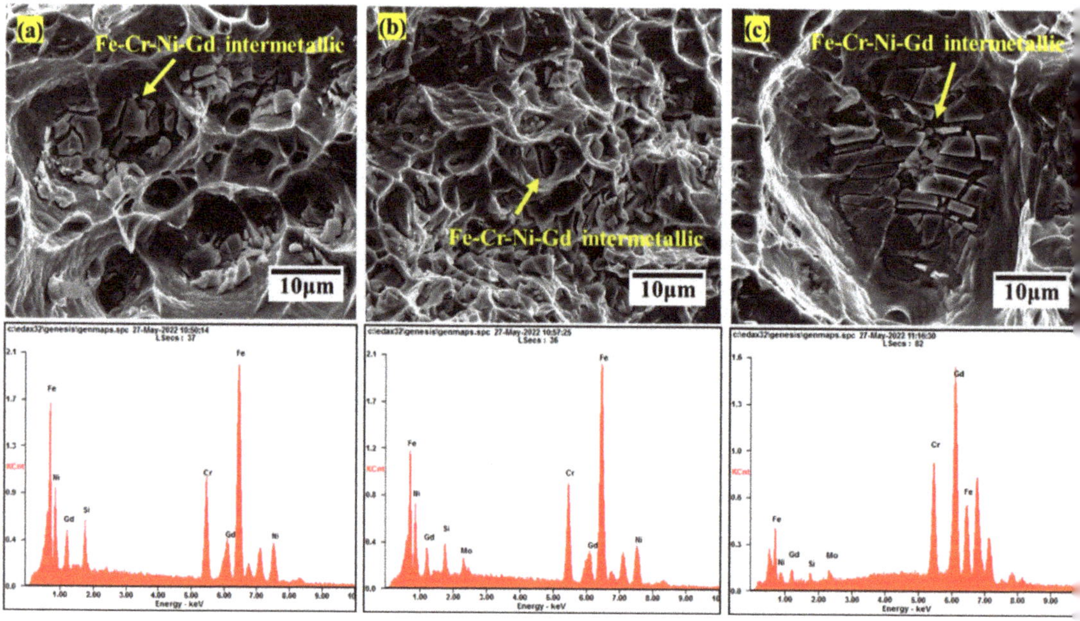

Figure 8. SEM images (**top**) of fracture surfaces and EDS analysis results (**bottom**) for the experimental alloys: (**a**) 304-Gd, (**b**) 316-Gd, and (**c**) 5A-Gd.

Table 4. Potentiodynamic polarization parameters of the experimental alloys.

Alloys	E_{corr} (mV vs. Ag/AgCl)	I_{corr} (μA/cm^2)	β_c (mV/decade)	β_a (mV/decade)	R_p (kΩ·cm^2)
304	−229.1	0.55	328.1	233.9	107.2
316	−180.6	0.19	128.8	109.9	133.1
5A	−220.1	0.38	145.6	375.5	119.7
304-Gd	−318.4	1.35	313.0	153.1	33.2
316-Gd	−299.1	2.68	387.5	501.0	35.4
5A-Gd	−245.8	4.84	498.4	305.8	17.0

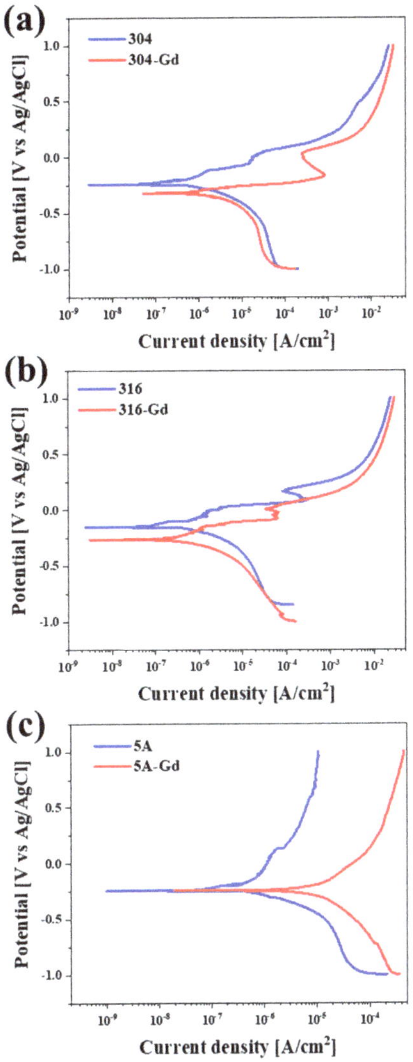

Figure 9. Potentiodynamic polarization curves for (**a**) 304 and 304-Gd, (**b**) 316 and 316-Gd, and (**c**) 5A and 5A-Gd in artificial seawater.

3.4. Thermal-Neutron-Absorbing Properties of the Alloys

Theoretical calculations of thermal neutron absorption in the fabricated alloys were conducted utilizing the Beer–Lambert Law. In Figure 10, we present the correlation between the thermal neutron transmission ratio (I/I_0) and varying Gd concentrations in fabricated steel alloys with different thicknesses of alloys. All the graphs exhibit a consistent pattern characterized by an exponential reduction in the neutron transmission ratio with increasing Gd concentrations in the steel-alloy-based thermal-neutron-absorbing material. Furthermore, with a thicker fabricated steel alloy, the thermal neutron transmission ratio decreased. Consequently, when the content of Gd in alloys was 2 wt%, nearly all neutrons were effectively shielded.

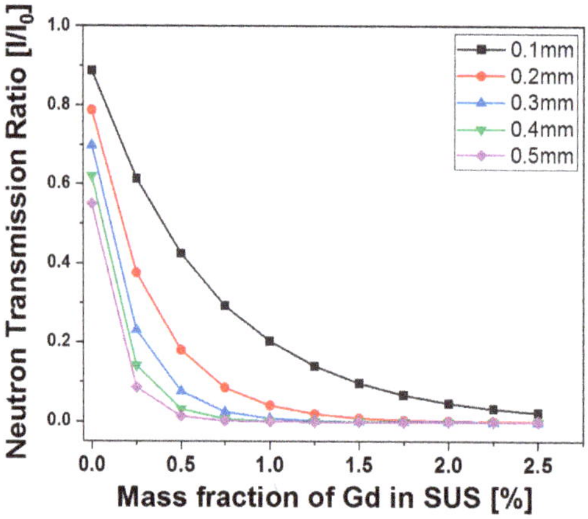

Figure 10. The relationship between the neutron transmission ratio and contents of Gd in various steels.

In our previous paper, the thermal neutron transmission ratio of the Fe alloy with 1 wt% of Gd was found to be similar to that of the Al composite with 30 vol% of boron carbide (B_4C), which is extensively researched as a thermal-neutron-absorbing and -shielding material [1]. It is important to highlight that even a small amount of Gd can effectively absorb thermal neutrons. Moreover, as shown in Figure 10, we observed that the neutron transmission ratio exponentially decreased with the increase in the thickness of the fabricated alloys and the Gd content in steel. It is widely recognized that the uniform distribution of nano-sized dispersoids within the matrix, which serve as anchoring points for dislocations and repositories for irradiation-induced defects, can notably enhance both the elevated mechanical properties and radiation resistance of alloys [18]. Consequently, the simultaneous improvement of mechanical properties and thermal-neutron-shielding performance of the alloy appears promising and was achieved through the uniform dispersion of Gd-based intermetallics in the matrix. Although the mechanical properties and corrosion resistance decreased in all conditions due to the strong oxidizing nature of Gd, which led to the formation of numerous Gd-based oxides, the results were still higher than those of commercial borated steel. These outcomes suggest that steel alloys with 2 wt% Gd show immense potential as thermal-neutron-absorbing or -shielding materials that are capable of surpassing the constraints faced by current thermal-neutron-absorbing and -shielding structural materials.

In recent years, rare earth elements have gained prominence in the realm of thermal neutron absorption due to their substantial neutron absorption cross-section. Notably, certain rare earth elements, exemplified by Gd, exhibit a notably higher thermal neutron absorption cross-section compared with boron. Consequently, these elements have evolved into pivotal constituents of shielding materials [32,33]. It becomes evident that an ideal formulation for thermal-neutron-shielding materials should encompass a combination of both base metals, such as Fe-based alloys and Al-based alloys, and heavy rare earth elements. These rare earth particles possess a distinctive capability to absorb low-energy neutrons and operate in concert to moderate higher-energy neutrons through inelastic scattering [34].

Enhancing the efficient utilization of rare earth elements and optimizing the design of shielding capabilities is imperative. This strategic approach serves as a foundational cornerstone for advancing the overall performance of materials, encompassing attributes like mechanical properties, functionalization, and heightened shielding efficacy. In this study, we used Gd as a thermal-neutron-shielding and -absorbing material, but it is possible to create materials with thermal-neutron-shielding ability using other rare earth metals as well. We plan to conduct further research on this. Additionally, enhancing the efficient utilization of rare earth elements and optimizing the design of shielding capabilities is imperative. This strategic approach serves as a foundational cornerstone for advancing the overall performance of materials, encompassing attributes like lightweight properties, functionalization, and heightened shielding efficacy.

4. Conclusions

In this study, Gd was added to commercial 304 and 316 austenitic and DSS 5A alloys to confirm its potential as a next-generation neutron absorber. The experimental alloys were manufactured using VAM, and 2 wt.% of Gd was added to 304, 316, and 5A commercial alloys. After a solution heat treatment, the microstructure, mechanical properties, and corrosion resistance were evaluated. The mechanical property evaluation results were compared with the minimum physical property requirements for borated stainless steel in ASTM standard A887. As a result of the microstructure observation, Gd was found to be uniformly distributed in the alloy as intermetallics. There was no segregation or clustering, and there were no coarse intermetallics. The Gd intermetallic peak was also not observed via XRD analysis. The tensile test results show that the ultimate tensile strength and yield strength increased by up to 11% after the addition of Gd. Although the elongation decreased by up to 48%, a significantly higher elongation value was observed compared with that of borated stainless steel. The Charpy impact test also showed a decrease in impact energy by up to 56% after the addition of Gd, but the impact energy was twice as high as that of borated stainless steel. As a result of analyzing the SEM fractographs, it was confirmed that brittle fracturing occurred in some areas after the Gd was added. Additionally, Gd intermetallics were present inside the dimple and showed a broken shape. Potentiodynamic polarization tests of the experimental alloys were performed, and the corrosion resistance deteriorated after Gd was added. The Rp values decreased by 69.01%, 74.41%, and 85.78% for 304, 316, and 5A, respectively, after Gd was added. The mechanical properties of the experimental alloys were confirmed to be superior to those of borated stainless steel and the neutron absorption capacity was also confirmed to be superior to that of the additive element.

Supplementary Materials: The following supporting information can be downloaded from https://www.mdpi.com/article/10.3390/met14010016/s1—Table S1: Neutron absorbing cross-sections of Fe and Gd.

Author Contributions: S.O., H.-D.J., R.J. and H.-J.K.; methodology, S.O., J.-H.A., D.H.C. and Y.C.; validation, J.-H.A. and H.-D.J.; investigation, R.J., H.-J.K. and Y.C.; data curation, R.J., H.-J.K. and Y.C.; formal analysis, S.O., H.L. and H.-D.J.; writing—original draft preparation, S.O. and H.L.; supervision, D.H.C. and H.-D.J.; project administration, Y.C. and H.-D.J.; funding acquisition. All authors have read and agreed to the published version of the manuscript.

Funding: This work was supported by The Catholic University of Korea, Research Fund, 2022; the National Research Foundation of Korea (NRF) grant funded by the Korean government (MSIT) (no. 2021R1A2C1091301); and the Technology Innovation program managed by the Ministry of Science and ICT and the Ministry of Trade, Industry, and Energy (MOTIE) (nos. 20010488 and 20019192).

Data Availability Statement: The data presented in this study are available on request from the corresponding author. The data are not publicly available due to privacy.

Conflicts of Interest: Author Younghwan Chu was employed by the company Fospia Co., Ltd.; Author Dae Hyun Choi was employed by the company SHINWOO CNC KOREA Co., LTD. The remaining authors declare that the research was conducted in the absence of any commercial or financial relationships that could be construed as a potential conflict of interest.

ferences

Lee, S.-W.; Ahn, J.-H.; Moon, B.-M.; Kim, D.; Oh, S.; Kim, Y.-J.; Jung, H.-D. Preliminary study on FeGd alloys as binary alloys and master alloys for potential spent nuclear fuel (SNF) application. *Mater. Des.* **2020**, *194*, 108906. [CrossRef]

Moon, B.-M.; Lee, S.-W.; Kim, M.-J.; Jung, S.-R.; Kim, Y.-J.; Jung, H.-D. The effect of gadolinium on the microstructures and Charpy impact properties of super duplex stainless steels. *Metals* **2018**, *8*, 474. [CrossRef]

3. Taylor, R.; Mathers, G.; Banford, A. The development of future options for aqueous recycling of spent nuclear fuels. *Prog. Nu Energy* **2023**, *164*, 104837. [CrossRef]
4. Qi, Z.; Yang, Z.; Li, J.; Guo, Y.; Yang, G.; Yu, Y.; Zhang, J. The advancement of neutron-shielding materials for the transportati and storage of spent nuclear fuel. *Materials* **2022**, *15*, 3255. [CrossRef] [PubMed]
5. Karley, D.; Shukla, S.K.; Rao, T.S. Microbiological assessment of spent nuclear fuel pools: An in-perspective review. *J. Envir Chem. Eng.* **2022**, *10*, 108050. [CrossRef]
6. Holdsworth, A.F.; Eccles, H.; Sharrad, C.A.; George, K. Spent Nuclear Fuel—Waste or Resource? The Potential of Strate Materials Recovery during Recycle for Sustainability and Advanced Waste Management. *Waste* **2023**, *1*, 249–263. [CrossRef]
7. Gan, B.; Liu, S.; He, Z.; Chen, F.; Niu, H.; Cheng, J.; Tan, B.; Yu, B. Research progress of metal-based shielding materials neutron and gamma rays. *Acta Metall. Sin. (Engl. Lett.)* **2021**, *34*, 1609–1617. [CrossRef]
8. Qi, Z.-D.; Yang, Z.; Yang, X.-G.; Wang, L.-Y.; Li, C.-Y.; Dai, Y. Performance study and optimal design of Gd/316L neutr absorbing material for Spent Nuclear Fuel transportation and storage. *Mater. Today Commun.* **2023**, *34*, 105342. [CrossRef]
9. Jung, Y.; Lee, M.; Kim, K.; Ahn, S. 10B (n, α) 7Li reaction-induced gas bubble formation in Al–B4C neutron absorber irradiated spent nuclear fuel pool. *J. Nucl. Mater.* **2020**, *533*, 152077. [CrossRef]
10. Jiang, L.; Xu, Z.; Fei, Y.; Zhang, Q.; Qiao, J.; Wu, G. The design of novel neutron shielding (Gd + B4C)/6061Al composites and properties after hot rolling. *Compos. Part B Eng.* **2019**, *168*, 183–194. [CrossRef]
11. Kim, M.J.; Kim, W.; Lee, D.; Lemaire, M.; Lee, H.-J.; Sohn, D.-S.; Kwon, H. Development of integral type spent fuel pool stora rack with gadolinium and europium-containing structure materials. *Ann. Nucl. Energy* **2019**, *130*, 107–117. [CrossRef]
12. Hur, D.H.; Jeon, S.-H.; Han, J.; Park, S.-Y.; Chun, Y.-B. Effect of gadolinium addition on the corrosion behavior and oxi properties of titanium in boric acid solution at 50 °C. *J. Mater. Res. Technol.* **2022**, *21*, 3051–3061. [CrossRef]
13. Xu, Z.; Jiang, L.; Zhang, Q.; Qiao, J.; Gong, D.; Wu, G. The design of a novel neutron shielding B4C/Al composite containing (*Mater. Des.* **2016**, *111*, 375–381. [CrossRef]
14. Cong, S.; Ran, G.; Li, Y.; Chen, Y. Ball-milling properties and sintering behavior of Al-based Gd2O3–W shielding materials us in spent-fuel storage. *Powder Technol.* **2020**, *369*, 127–136. [CrossRef]
15. Lister, T.; Mizia, R.; Pinhero, P.; Trowbridge, T.; Delezene-Briggs, K. Studies of the corrosion properties of Ni-Cr-Mo-(neutron-absorbing alloys. *Corrosion* **2005**, *61*, 706–717. [CrossRef]
16. Mizia, R.E.; Lister, T.E.; Pinhero, P.J.; Trowbridge, T.L.; Hurt, W.L.; Robino, C.V.; Stephens, J.J., Jr.; DuPont, J.N. Developme and testing of an advanced neutron-absorbing gadolinium alloy for spent nuclear fuel storage. *Nucl. Technol.* **2006**, *155*, 133–1 [CrossRef]
17. Wang, W.; Zhang, J.; Wan, S.; Zhang, T. Design, fabrication and comprehensive properties of the novel thermal neutron shieldi Gd/316L composites. *Fusion Eng. Des.* **2021**, *171*, 112566. [CrossRef]
18. Yang, X.; Song, L.; Chang, B.; Yang, Q.; Mao, X.; Huang, Q. Development of Gd-Si-O dispersed 316L stainless steel for improvi neutron shielding performance. *Nucl. Mater. Energy* **2020**, *23*, 100739. [CrossRef]
19. Wang, H.; Wang, T.; Peng, J. Effect of gadolinium addition on microstructure, mechanical properties and corrosion resistance 316L austenitic stainless steels. *Phys. Met. Metallogr.* **2021**, *122*, 1640–1647. [CrossRef]
20. Baik, Y.; Choi, Y.; Moon, B.M.; Sohn, D.S.; Bogdanov, S.G.; Pirogov, A.N. Effect of gadolinium addition on the corrosion, wear, a neutron absorbing behaviors of duplex stainless steel sheet. *Phys. Met. Metallogr.* **2015**, *116*, 1135–1142. [CrossRef]
21. Ahn, J.-H.; Jung, H.-D.; Im, J.-H.; Jung, K.H.; Moon, B.-M. Influence of the addition of gadolinium on the microstructure a mechanical properties of duplex stainless steel. *Mater. Sci. Eng. A* **2016**, *658*, 255–262. [CrossRef]
22. Jung, M.Y.; Baik, Y.; Choi, Y.; Sohn, D.-S. Corrosion and mechanical properties of hot-rolled 0.5% Gd-0.8% B-stainless steels ir simulated nuclear waste treatment solution. *Nucl. Eng. Technol.* **2019**, *51*, 207–213. [CrossRef]
23. Schmidt, M.; Del Corso, G.; Klankowski, K.; Lherbier, L.; Novotnak, D. Review of the Development and Testing of a N Family of Boron and Gadolinium-bearing Dual Thermal Neutron Absorbing Alloys-13026. In Proceedings of the WM2013: Wa Management Conference: International Collaboration and Continuous Improvement, Phoenix, AZ, USA, 24–28 February 20 pp. 24–28.
24. Saeed, A. Developed borated austenitic stainless steel alloys as nuclear reactor control rods. *Nucl. Eng. Des.* **2023**, *413*, 1125 [CrossRef]
25. Kang, M.-H.; Cheon, K.-H.; Jo, K.-I.; Ahn, J.-H.; Kim, H.-E.; Jung, H.-D.; Jang, T.-S. An asymmetric surface coating strategy improved corrosion resistance and vascular compatibility of magnesium alloy stents. *Mater. Des.* **2020**, *196*, 109182. [CrossRe
26. Lister, T.; Mizia, R.E.; Erickson, A.; Matteson, B. General and Localized Corrosion of Borated Stainless Steels. In Proceedings the NACE CORROSION, New Orleans, LA, USA, 16–20 March 2008; p. NACE-08590.
27. Lister, T.; Mizia, R.; Erickson, A.; Trowbridge, T. *Electrochemical Corrosion Testing of Neutron Absorber Materials*; Idaho National L (INL): Idaho Falls, ID, USA, 2007.
28. Robino, C.; Cieslak, M. High-temperature metallurgy of advanced borated stainless steels. *Metall. Mater. Trans. A* **1995**, 1673–1685. [CrossRef]
29. He, J. *The Effect of Boron Content, Temperature, and Neutron Fluence on the Mechanical Properties of the Modified Type 304 Stainless Ste with Boron*; The Pennsylvania State University: State College, PA, USA, 1995.
30. Wan, S.; Wang, W.; Chen, H.; Zhou, J.; Zhang, Y.; Liu, R.; Feng, R. 155/157Gd areal density: A model for design and fabrication Gd2O3/316L novel neutron shielding composites. *Vacuum* **2020**, *176*, 109304. [CrossRef]

Ha, H.-Y.; Lee, T.-H.; Jo, H.-H.; Lee, J.; Jang, J.H. Corrosion behavior of (Fe, Ni)–Gd intermetallic compounds in FeNi-based neutron-absorbing steels. *J. Nucl. Mater.* **2023**, *578*, 154367. [CrossRef]
Florez, R.; Loaiza, A.; Giraldo, C.H.C.; Colorado, H.A. Calcium silicate phosphate cement with samarium oxide additions for neutron shielding applications in nuclear industry. *Prog. Nucl. Energy* **2021**, *133*, 103650. [CrossRef]
Castley, D.; Goodwin, C.; Liu, J. Computational and experimental comparison of boron carbide, gadolinium oxide, samarium oxide, and graphene platelets as additives for a neutron shield. *Radiat. Phys. Chem.* **2019**, *165*, 108435. [CrossRef]
Jing, H.; Geng, L.; Qiu, S.; Zou, H.; Liang, M.; Deng, D. Research progress of rare earth composite shielding materials. *J. Rare Earths* **2023**, *41*, 32–41. [CrossRef]

claimer/Publisher's Note: The statements, opinions and data contained in all publications are solely those of the individual ıor(s) and contributor(s) and not of MDPI and/or the editor(s). MDPI and/or the editor(s) disclaim responsibility for any injury to ple or property resulting from any ideas, methods, instructions or products referred to in the content.

metals

Article

Effect of Double-Quenching on the Hardness and Toughness of a Wear-Resistant Steel

Jingliang Wang [1,2], Rongtao Qian [1], Song Huang [1,3] and Chengjia Shang [1,2,3,*]

1 Collaborative Innovation Center of Steel Technology, University of Science and Technology Beijing, Beijing 100083, China
2 Yangjiang Branch, Guangdong Laboratory for Materials Science and Technology (Yangjiang Advanced Alloys Laboratory), Yangjiang 529500, China
3 State Key Laboratory of Metal Materials for Marine Equipment and Application, Anshan 114021, China
* Correspondence: cjshang@ustb.edu.cn

Abstract: Martensitic/bainitic wear-resistant steels are widely used in civilian industry, where a good combination of strength and toughness is required. In the present study, a double-quenching process was applied and compared to the conventional single-quenching process. The microhardness and ductile–brittle transition temperature were measured, and the microstructure was characterized with scanning electron microscopy and electron backscatter diffraction (EBSD) technique. It was found that the double-quenching process refined the prior austenite grain size by 43% and simultaneously improved the toughness and hardness. The ductile-to-brittle transition temperature was decreased from −77 °C to −90 °C, and the hardness was increased by 8%. Based on the EBSD data, a detailed analysis of the grain boundary distribution was performed using a recently developed machine learning model. Unlike what was found in previous studies, for the studied wear-resistant steel, the refinement of the prior austenite grain did not increase the block boundary density while increasing the high-angle packet boundary density. As a result, the total density of the high-angle grain boundaries in the double-quenched specimen was not improved compared to the single-quenched specimen. Further inspection suggested that it is the prior austenite grain boundaries and high-angle packet boundaries that contribute to the hardness and toughness, and the key factors that determine their effectiveness are the high misorientation angle between the {110} slip planes and the high slip transmission factor.

Keywords: martensitic and bainitic steels; grain boundaries; toughness; crystallography; high-strength low-alloy steels; wear-resistant steels; heat treatment

1. Introduction

Martensitic wear-resistant steels are widely employed in civil constructions, where a combination of high strength and high toughness is of great importance. While as-quenched martensite provides a sufficient hardness, a subsequent tempering process is needed to achieve the required toughness. Multiple studies have been carried out to investigate the effects of quenching and tempering conditions on the mechanical properties of wear-resistant steels. While the impact toughness can be improved by an appropriate increase in the tempering temperature, temper embrittlement takes place when the tempering temperature exceeds a critical value. Meanwhile, the hardness decreases constantly with increasing tempering temperature [1–3]. On the other hand, simultaneous improvements in strength and toughness can be achieved by grain refinement via lowering the quenching temperature. Haiko et al. altered the reaustenitization (quenching) temperatures of an as-quenched 500 HB wear-resistant steel and found that a finer prior austenite grain (PAG) size led to a significant improvement in low-temperature impact toughness and a slight increase in hardness [4]. Cyclic quenching (reaustenitizing and quenching two or more times) has also proved to be an effective way to refine the grain size and improve the comprehensive properties of a wide range of steels [5–10]. However, so far, studies on the cyclic quenching of martensitic/bainitic wear-resistant steels are still scarce.

In the martensitic/bainitic microstructure, a PAG is divided into several packets. Each packet consists of laths with the same habit plane and is further divided into blocks consisting of laths with similar orientations [11]. Corresponding to the hierarchical microstructure, there are different types of grain boundaries. While the positive effect of grain refinement on the mechanical properties of steels is commonly described by the Hall–Petch relationship, the 'effective' grain size, or in other words, the effective grain boundary governing the relevant properties in a martensitic/bainitic microstructure, is still under debate. Concerning the toughness, Wang et al. proposed that packet boundaries strongly hindered fracture propagation and packets acted as the effective microstructure unit for cleavage [8]. It was also reported that the finer packet size or higher density of packet boundaries decreased the ductile-to-brittle transition temperature (DBTT) [12,13]. Zhang et al. argued that block boundaries dominate the high-angle boundaries and blocks should be the minimum structure unit controlling the toughness [14]. As reported in Ref. [15], there were many studies that supported that the block boundary is the key microstructural feature dominating the toughness and strength. Morito et al. analyzed the yield strengths of low-carbon martensitic steels with respect to the packet and block sizes, respectively, and the results suggested that the block size is the key structural parameter for the strength [16]. Wang et al. [17] also found a near-linear relationship between the block boundary density and the hardness of two bainitic/martensitic steels.

The overall misorientation angle (OMA), or simply the misorientation angle, was usually used as the criterion to define high-angle or low-angle grain boundaries. It has long been recognized that the high-angle (OMA) grain boundaries are effective in impeding crack propagation. Guo et al proposed, more specifically, that the crystallographic feature that governs transgranular fracture is the {100} cleavage plane, while the dislocation plasticity is governed by the {110} slip plane [18]. The specific misorientation angle (SMA), as opposed to the OMA, was thus proposed to describe the misorientation relationship (OR) [13,19,20]. The SMA was defined as the angle between specific crystallographic planes of two neighboring sub-volumes. The SMAs between {100} cleavage planes ({100}-SMA) and between {110} slip planes ({110}-SMA) were considered to be relevant to the cleavage fracture and dislocation gliding, respectively. In low-carbon martensitic/bainitic steels, the parent austenite phase and product phase often hold the Kurdjumov–Sachs (K–S) relationship, which is characterized by the parallelism between close-packed planes and directions in the two phases: $\{111\}_{\gamma}||\{110\}_{\alpha'}$ and $<110>_{\gamma}||<111>_{\alpha'}$. Due to symmetry, for the K–S relationship there are four possible habit planes, $(111)_{\gamma}$, $(\bar{1}11)_{\gamma}$, $(1\bar{1}1)_{\gamma}$, and $(11\bar{1})_{\gamma}$, i.e., four packets within a PAG, and there are six variants, i.e., six blocks within each packet. Theoretically, there are three possible inter-variant OMAs, 10.53°, 49.47°, and 60°, within a packet [11], and in practice small deviations from the theoretical OR are usually observed. In the following, the inter-variant boundaries within a packet with an OMA smaller and greater than 15° are referred to as sub-block boundaries and block boundaries, respectively. The packet boundaries with OMA values smaller and greater than 40° are referred to as low-angle and high-angle packet boundaries, respectively. In summary, each inter-variant grain boundary within a PAG holds a theoretically fixed OR and can be categorized as a packet or block/sub-block boundary, while there are no defined ORs for PAG boundaries (PAGBs). A machine learning model was developed to distinguish different types of grain boundaries based on their features in four parameters: OMA, {100}-SMA, {110}-SMA, and {111}-SMA [13]. For example, a sub-block or block boundary must have a {110}-SMA close to zero due to the K–S relationship, while a packet boundary usually has a larger {110}-SMA.

In the present study, the effect of the double-quenching process on a wear-resistant steel was examined. The mechanical properties and microstructure of a double-quenched specimen were compared with those of a single-quenched specimen. Special attention was paid to the distribution of different types of grain boundaries, and the key microstructural feature that affects the mechanical properties of the martensitic/bainitic wear-resistant steel is discussed.

2. Materials and Methods

The as-rolled NM400 wear-resistant steel plate was received from the industry with the chemical composition listed in Table 1. Two groups of specimens were prepared. The first group was austenitized at 900 °C and then quenched in oil; the process was repeated for the second group, as illustrated in Figure 1. These specimens are denoted as single-quenched (SQ900) and double-quenched (DQ900), respectively. After the heat treatment, the specimens were machined into 10 mm × 10 mm × 55 mm V-notched Charpy test specimens following the ASTM standard, and instrumented Charpy impact tests were performed at various temperatures ranging from −120 °C to −40 °C. Three tests were performed for each condition. Vickers hardness tests were performed on the heat-treated specimens at 1 kgf, and the microstructure was characterized with a scanning electron microscope (SEM) and electron backscatter diffraction (EBSD). A TESCAN MIRA3 SEM system equipped with an Oxford Instruments Symmetry EBSD detector was employed for the SEM and EBSD analyses. For SEM imaging, the specimens were prepared by the standard metallographic method, including grinding with silicon carbide sandpapers and polishing with diamond polishing paste, followed by etching with 4% nital. For EBSD measurements, the specimens were electropolished after the mechanical polishing using a solution with 10% perchloric acid, 5% glycerol, and 85% ethanol.

Table 1. Chemical composition of the studied steel.

Element	C	Si	Mn	Cr + Mo	Nb + Ti	B	Fe
mass%	≤0.20	≤0.30	≤1.10	0.7–0.9	≤0.04	≤0.002	Bal.

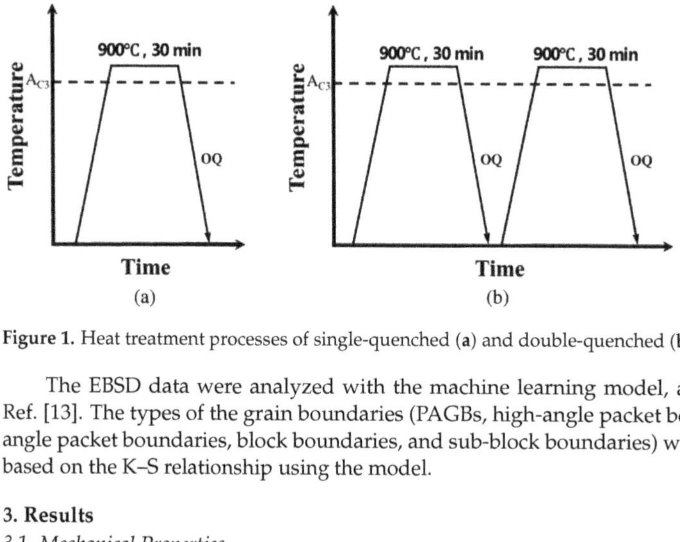

Figure 1. Heat treatment processes of single-quenched (**a**) and double-quenched (**b**) specimens.

The EBSD data were analyzed with the machine learning model, as presented in Ref. [13]. The types of the grain boundaries (PAGBs, high-angle packet boundaries, low-angle packet boundaries, block boundaries, and sub-block boundaries) were determined based on the K–S relationship using the model.

3. Results

3.1. Mechanical Properties

Figure 2a shows the total absorbed energy as a function of the temperature at which the Charpy test was performed. The DBTT was shifted from −77 °C to −90 °C by the extra quenching process, which suggests an improvement in the toughness. In contrast to the tempering process, where the increase in toughness is always associated with a degradation in hardness, the double-quenching process also led to a slight improvement in hardness: the microhardness increased from 379.7 HV to 408.5 HV (Figure 2b), with both meeting the requirement for the hardness of an NM400 steel.

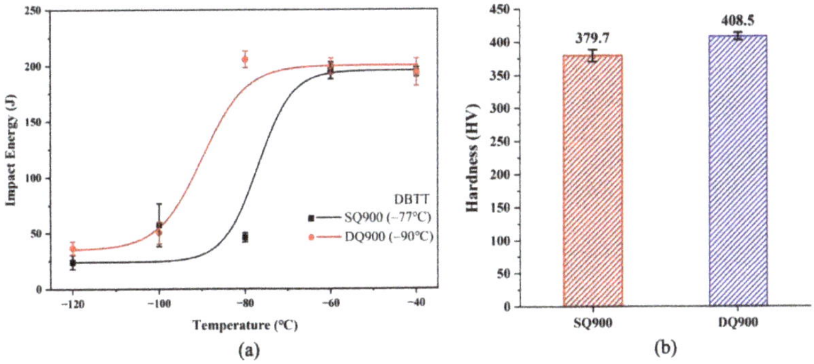

Figure 2. Impact energy–temperature curves (**a**) and microhardness (**b**) of the single-quenched and double-quenched specimens.

Unlike conventional Charpy impact tests, which only give the total absorbed energy, instrumented Charpy tests also record the impact load/energy–displacement curves. Figure 3 displays the curves measured at −80 °C, which is between the DBTT of the single-quenched and double-quenched specimens. The curves can be divided into two parts: the crack initiation stage and the crack propagation stage. The cracks were initiated at the maximum load (F_m), and beyond F_m the crack propagated steadily until a critical crack length was reached. After that, unstable crack propagation took place, manifested as a sudden drop in the impact load. The absorbed energy at F_m was the crack initiation energy (CIE), as illustrated in Figure 3, and the difference between the total absorbed energy and CIE was the crack propagation energy (CPE). The CIEs for the single-quenched and double-quenched specimens were quite similar (34 J and 38 J, respectively), and it was the distinct CPEs (16 J and 176 J, respectively) that accounted for the remarkably different total impact energies (50 J and 214 J, respectively) at −80 °C. It is clear that the double-quenched specimen underwent a longer ductile propagation [21] and exhibited a much better plasticity.

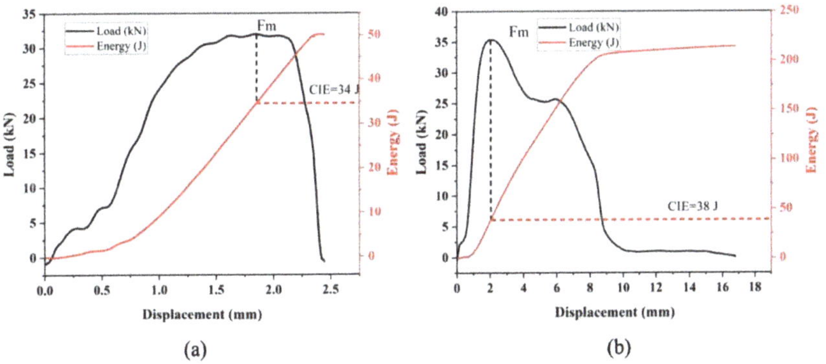

Figure 3. The impact load (energy)–displacement curves of the single-quenched (**a**) and double-quenched specimens (**b**) measured at −80 °C.

3.2. Microstructure Characterization

SEM images of the heat-treated specimens are shown in Figure 4. Both single-quenched and double-quenched specimens exhibited granular bainitic microstructures with fine, dispersed martensite/austenite islands, resembling that of a water-quenched and tempered NM400 steel with a similar chemical composition [22]. The PAGBs could be visualized, as the arrows indicate. It can be seen that the double-quenched specimen had a much smaller prior austenite grain size than the single-quenched specimen. As reported in the literature, the PAG size can be effectively refined by cyclic quenching due to more nuclei for austenite reversion generated from the preceding quenching [5,7,23].

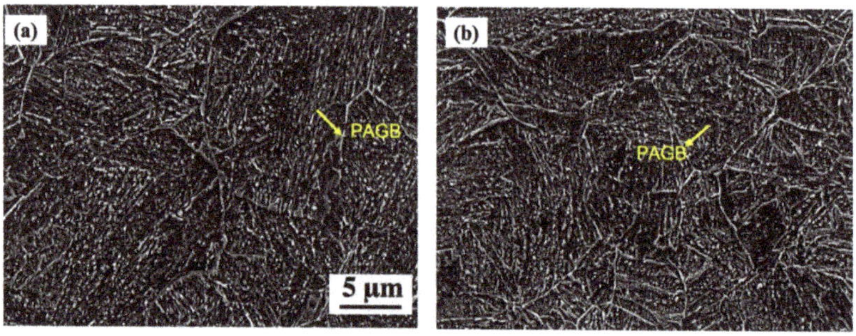

Figure 4. SEM images of the single-quenched (**a**) and double-quenched specimens (**b**).

Figure 5a presents the EBSD band contrast (BC) maps overlapped with grain boundaries shown as lines colored according to their OMAs. In Figure 5b, the grain boundaries are colored according to their types, with PAGBs shown separately in Figure 5c for the convenience of visualization. The PAG size was measured by the intercept method based on three sets of EBSD data for each specimen, and the average PAG size decreased from 12.7 ± 2.6 μm to 7.3 ± 0.6 μm by the double-quenching process. It can clearly be seen that the double-quenching process effectively refined and homogenized the PAG size. However, from Figure 5a,b, it seems that the total density of high-angle grain boundaries (with OMA > 45°) was reduced by the double-quenching process, which was confirmed by the quantitative analysis shown in Figure 6: double-quenching led to a higher PAGB density, corresponding to the decreased PAG size and a higher high-angle packet boundary density. However, the double-quenched specimen showed a much smaller block boundary density compared to the single-quenched specimen. As the block boundaries accounted for the majority of the high-angle grain boundaries, the total density of the high-angle grain boundaries of the double-quenched specimens was smaller than that of the single-quenched specimen.

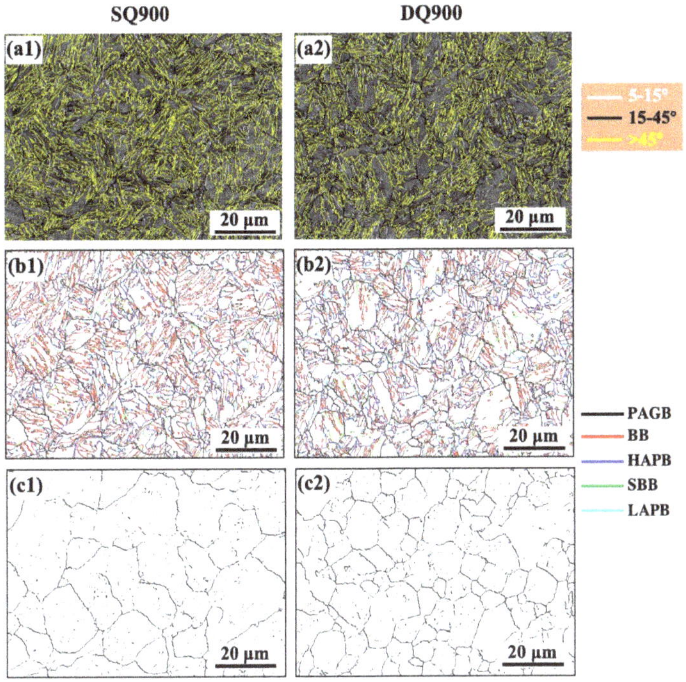

Figure 5. (**a**) EBSD band contrast map overlapped with grain boundaries, where lines of different colors represent different overall misorientation angles; (**b**) Grain boundary maps, where lines of different colors represent different types (PAGB: prior austenite grain boundary, HAPG: high-angle packet boundary, BB: block boundary, LAPG: low-angle packet boundary, SBB: sub-block boundary); (**c**) Prior austenite grain boundaries. The **left** and **right** columns are maps for the single-quenched and double-quenched specimens, respectively.

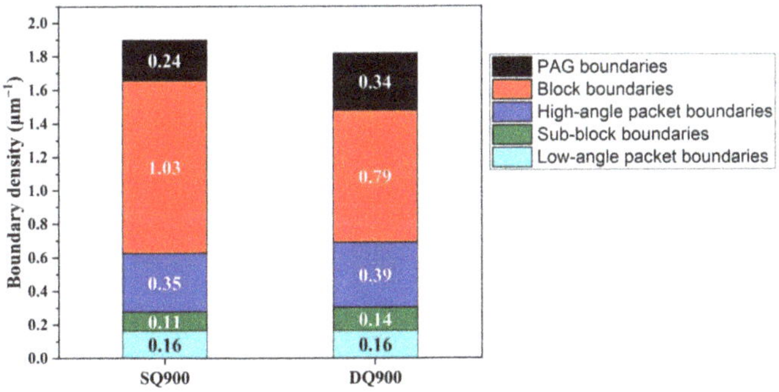

Figure 6. Density of grain boundaries of different types in the single-quenched and double-quenched specimens, quantified based on the EBSD data shown in Figure 5b.

4. Discussion

4.1. Effective Grain Boundaries for DBTT

The refinement of PAG size was reported to decrease both the packet and block sizes in lath martensite structures after quenching [5,24]. However, our results suggested that the double-quenched specimen showed a lower block boundary density compared to the single-quenched specimen. Our finding was consistent with Wang et al.'s work [25], where smaller PAGs were shown to have lower block boundary densities. The results imply that the block boundaries are not responsible for the improvement in the toughness and hardness. It is thus inferred that although both high-angle packet and block boundaries have a high OMA, their roles as barriers are quite different. The {110}-SMA was considered to control the plastic deformation because {110} planes are primary slip planes. Therefore, the grain boundary with a high {110}-SMA defines the effective grain within which the free slip can occur. As seen in Figure 3, the lowering of DBTT by the double-quenching process was owing to the high plastic deformation. Therefore, we looked into the {110}-SMA distribution in the two specimens, as shown in Figure 7a. Comparing Figure 7a to Figure 5b, we can see that the high {110}-SMA grain boundaries are mostly PAGBs and high-angle packet boundaries, which can be seen more clearly from the quantitative analysis (Figure 7b). The {110}-SMA of block boundaries are mostly below 5°, while those of high-angle packet boundaries are mostly between 5° and 15°. The majority of PAGBs also had {110}-SMAs greater than 5°. As a result, although the total density of high-angle (OMA) grain boundaries in the double-quenched specimen was smaller than that of the single-quenched specimen, the total density of the high-{110}-SMA grain boundaries was higher, as shown in Figure 8. The high density of PAGBs and high-angle packet boundaries in the double-quenched specimen would contribute effectively to the resistance to dislocation gliding. The high plastic deformation of the double-quenched specimen thus seems counterintuitive. However, it was also reported in other works that the refinement of PAG and packet size decreased the DBTT [12,13], consistent with our finding. A plausible explanation is that the fine effective grains for slip decentralized the stress concentration at the crack tip, so the unstable crack propagation was delayed. This was confirmed in Figure 3, where one can see a prolonged crack propagation stage for the double-quenched specimen.

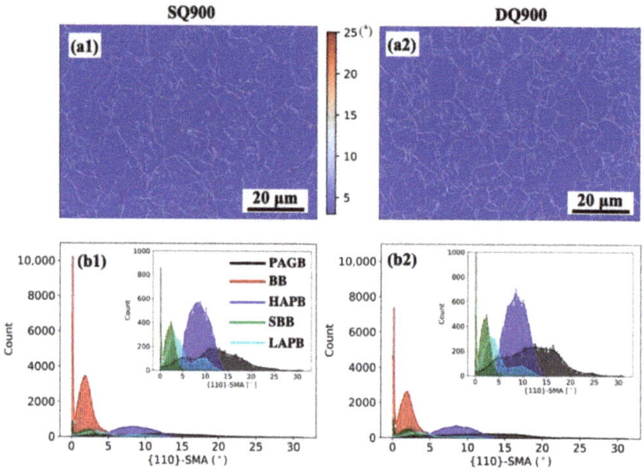

Figure 7. A {110}-SMA distribution map (**a**) and {110}-SMAs of different types of grain boundaries (**b**) of single-quenched (**left** column) and double-quenched (**right** column) specimens (PAGB: prior austenite grain boundary, HAPG: high-angle packet boundary, BB: block boundary, LAPG: low-angle packet boundary, SBB: sub-block boundary).

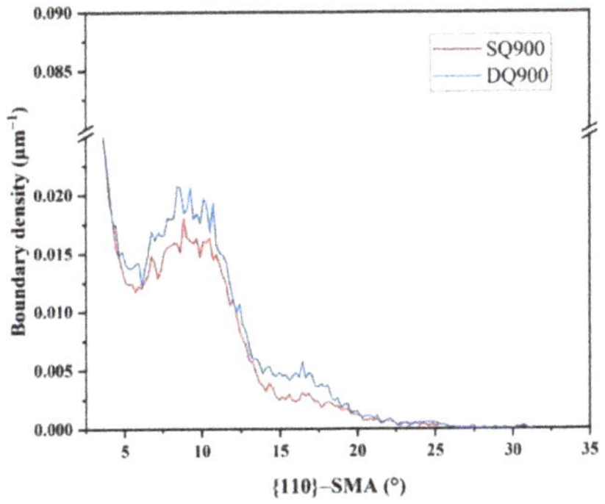

Figure 8. Density of boundaries with different {110}-SMAs.

4.2. Effective Grain Boundaries for Hardness

In previous studies, block boundaries were considered to be most relevant to the hardness [17]. In the present study, however, the double-quenched specimen with a lower block boundary density had a slightly higher hardness. The hardness reflects the resistance against localized plastic deformation, i.e., dislocation movement [26]. Therefore, the concept of {110}-SMA seems to also work for the hardness. In that sense, PAGBs and high-angle packet boundaries also contributed to the hardness.

Apart from {110}-SMA, we also calculated the slip transmission factor m′, which was defined as $m' = \cos(\varphi)\cos(\kappa)$, where φ is the angle between the slip directions in neighboring crystallographic units ('grains') and κ is the angle between normal to the slip planes of neighboring grains. The slip transmission factor is thus a geometric factor to measure the ability of slip transfer between neighboring grains without losing the coherency at the grain boundary [27]. The slip transmission factor between different variants was reported to be closely related to the plasticity and strength/hardness [28,29]. By definition, the higher the m′, the more transparent a grain boundary for a specific slip system. In this study, we took the N value, the maximum m′ of all slip systems in the adjacent sub-volumes, to indicate the ease of slip transmission of a grain boundary. In other words, the grain boundary with a larger 1-N value is more resistant to slip transmission. The distribution maps of 1-N values are shown in Figure 9a, and the quantitative results are plotted in Figure 9b. In contrast to the {110}-SMA, the 1-N value of the high-angle packet boundary shows a bimodal distribution, while it is still the PAGB and high-angle packet boundary that contribute most to the high 1-N values. Therefore, the double-quenched specimen also showed a higher 1-N value than the single-quenched specimen (Figure 10), i.e., a higher resistance to plastic deformation and hence higher hardness.

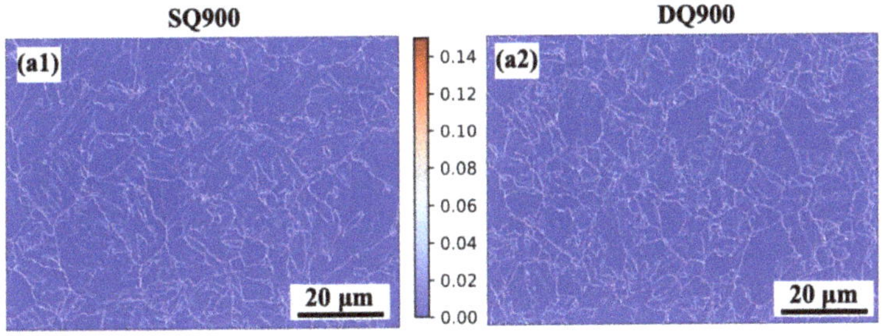

Figure 9. *Cont.*

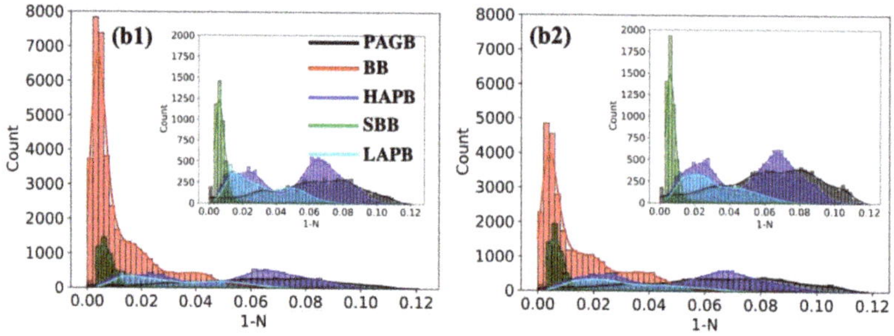

Figure 9. The 1-N value distribution map (**a**) and 1-N values of different types of grain boundaries (**b**) of single-quenched (**left** column) and double-quenched (**right** column) specimens (PAGB: prior austenite grain boundary, HAPG: high-angle packet boundary, BB: block boundary, LAPG: low-angle packet boundary, SBB: sub-block boundary).

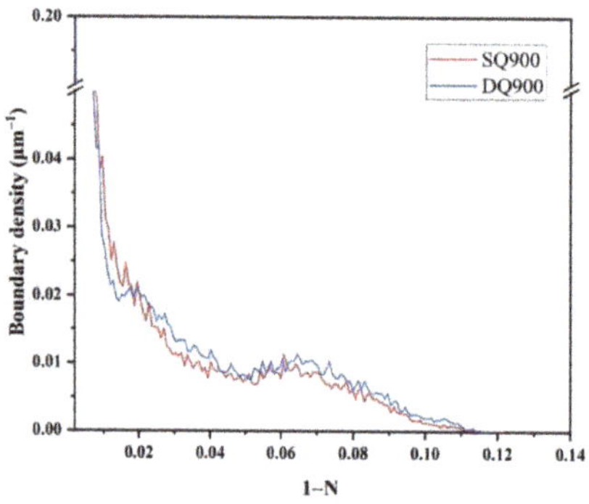

Figure 10. Density of boundaries with different 1-N values.

5. Conclusions

Compared to the conventional single-quenching process, the application of double-quenching to the wear-resistant steel refined the PAG size significantly and improved both the hardness and impact toughness. The quantitative analysis of the grain boundaries indicated that the double-quenched specimen had a higher PAGB and high-angle packet boundary density, while it showed a lower block boundary density. Further inspection suggested that both the PAGB and high-angle packet boundary showed high {110}-SMA and 1-N values compared to the other types of grain boundaries. Therefore, they were more effective in hindering dislocation movement, which reduced the stress concentration at the crack tip during the Charpy test. Meanwhile, the restriction of dislocation movement also contributed positively to the hardness.

Author Contributions: Conceptualization, C.S.; methodology, C.S. and J.W.; formal analysis, J.W. and R.Q.; investigation, R.Q. and S.H.; resources, S.H., J.W. and C.S.; data curation, J.W. and R.Q; writing—original draft preparation, J.W.; writing—review and editing, J.W.; visualization, R.Q. and J.W.; supervision, J.W. and C.S.; funding acquisition, C.S. and J.W. All authors have read and agreed to the published version of the manuscript.

Funding: This research was funded by the National Natural Science Foundation of China (grant number 52271089) and the Fundamental Research Funds for the Central Universities (grant number FRF-IC-20-04).

Data Availability Statement: The data presented in this study are available on request from the corresponding author.

Acknowledgments: The authors are grateful to Jingxiao Zhao for his help with the grain boundary analysis.

Conflicts of Interest: The authors declare no conflict of interest.

erences

Wen, E.; Song, R.; Xiong, W. Effect of tempering temperature on microstructures and wear behavior of a 500 HB grade wear-resistant steel. *Metals* **2019**, *9*, 45. [CrossRef]
Cao, Y.; Wang, Z.; Kang, J.; Wu, D.; Wang, G. Effects of tempering temperature and Mo/Ni on microstructures and properties of lath martensitic wear-resistant steels. *J. Iron Steel Res. Int.* **2013**, *20*, 70–75. [CrossRef]
Fu, H.; Xiao, Q.; Fu, H. Heat treatment of multi-element low alloy wear-resistant steel. *Mater. Sci. Eng. A* **2005**, *396*, 206–212. [CrossRef]
Haiko, O.; Javaheri, V.; Valtonen, K.; Kaijalainen, A.; Hannula, J.; Kömi, J. Effect of prior austenite grain size on the abrasive wear resistance of ultra-high strength martensitic steels. *Wear* **2020**, *454*, 203336. [CrossRef]
Furuhara, T.; Kikumoto, K.; Saito, H.; Sekine, T.; Ogawa, T.; Morito, S.; Maki, T. Phase transformation from fine-grained austenite. *ISIJ Int.* **2008**, *48*, 1038–1045. [CrossRef]
Kamada, Y.; Ohnishi, K.; Bessyo, K. Effect of double-quench processing on toughness of heavy-gauge ultrahigh strength steel plate. *J. Mater. Sci. Lett.* **1991**, *10*, 598–601. [CrossRef]
Lee, K.; Hong, S.; Kang, Y.; Yoon, H.; Kang, S. Grain refinement in bearing steels using a double-quenching heat-treatment process. *Int. J. Automot. Technol.* **2009**, *10*, 697–702. [CrossRef]
Wang, C.; Wang, M.; Shi, J.; Hui, W.; Dong, H. Effect of microstructural refinement on the toughness of low carbon martensitic steel. *Scripta Mater.* **2008**, *58*, 492–495. [CrossRef]
Sanij, M.K.; Banadkouki, S.G.; Mashreghi, A.; Moshrefifar, M. The effect of single and double quenching and tempering heat treatments on the microstructure and mechanical properties of AISI 4140 steel. *Mater. Des.* **2012**, *42*, 339–346. [CrossRef]
Cao, Z.; Shi, Z.; Yu, F.; Sugimoto, K.-i.; Cao, W.; Weng, Y. Effects of double quenching on fatigue properties of high carbon bearing steel with extra-high purity. *Int. J. Fatigue* **2019**, *128*, 105176. [CrossRef]
Morito, S.; Tanaka, H.; Konishi, R.; Furuhara, T.; Maki, T. The morphology and crystallography of lath martensite in Fe-C alloys. *Acta Mater.* **2003**, *51*, 1789–1799. [CrossRef]
Liu, D.; Luo, M.; Cheng, B.; Cao, R.; Chen, J. Microstructural evolution and ductile-to-brittle transition in a low-carbon MnCrMoNiCu heavy plate steel. *Metall. Mater. Trans. A* **2018**, *49*, 4918–4936. [CrossRef]
Li, X.; Zhao, J.; Cong, J.; Misra, R.; Wang, X.; Wang, X.; Shang, C. Machine learning guided automatic recognition of crystal boundaries in bainitic/martensitic alloy and relationship between boundary types and ductile-to-brittle transition behavior. *J. Mater. Sci. Technol.* **2021**, *84*, 49–58. [CrossRef]
Zhang, C.; Wang, Q.; Ren, J.; Li, R.; Wang, M.; Zhang, F.; Sun, K. Effect of martensitic morphology on mechanical properties of an as-quenched and tempered 25CrMo48V steel. *Mater. Sci. Eng. A* **2012**, *534*, 339–346. [CrossRef]
Morris Jr, J.W. Comments on the microstructure and properties of ultrafine grained steel. *ISIJ Int.* **2008**, *48*, 1063–1070. [CrossRef]
Morito, S.; Yoshida, H.; Maki, T.; Huang, X. Effect of block size on the strength of lath martensite in low carbon steels. *Mater. Sci. Eng. A* **2006**, *438*, 237–240. [CrossRef]
Wang, J.; Hong, H.; Huang, A.; Yang, X.; Qian, R.; Shang, C. New insight into the relationship between grain boundaries and hardness in bainitic/martensitic steels from the crystallographic perspective. *Mater. Lett.* **2022**, *308*, 131105. [CrossRef]
Guo, Z.; Lee, C.; Morris Jr, J. On coherent transformations in steel. *Acta Mater.* **2004**, *52*, 5511–5518. [CrossRef]
Zhao, J.; Li, X.; Wang, X.; Liu, S.; Wang, X.; Shang, C. Distribution feature of specific misorientation angle in a bainitic steel. *Mater. Charact.* **2021**, *172*, 110874. [CrossRef]
Li, X.; Zhao, J.; Wang, J.; Wang, X.; Liu, S.; Shang, C. Effect of boundaries on toughness in high-strength low-alloy steels from the view of crystallographic misorientation. *Mater. Lett.* **2020**, *259*, 126841. [CrossRef]
Song, H.; Li, C.; Lan, L.; Zhao, D.; Wang, G. Impact toughness of an NM400 wear-resistant steel. *J. Iron Steel Res. Int.* **2013**, *20*, 72–77. [CrossRef]
Huang, S.; Yu, Y.; Wang, Z.; Su, S.; Chen, K.; Yuan, S.; Xie, Z.; Shang, C. Crystallographic insights into the role of nickel on hardenability of wear-resistant steels. *Mater. Lett.* **2022**, *306*, 130961. [CrossRef]
Yuan, S.; Xie, Z.; Wang, J.; Zhu, L.; Yan, L.; Shang, C.; Misra, R. Effect of heterogeneous microstructure on refining austenite grain size in low alloy heavy-gage plate. *Metals* **2020**, *10*, 132. [CrossRef]
Li, X.; Lu, G.; Wang, Q.; Zhao, J.; Xie, Z.; Misra, R.D.K.; Shang, C. The Effects of Prior Austenite Grain Refinement on Strength and Toughness of High-Strength Low-Alloy Steel. *Metals* **2021**, *12*, 28. [CrossRef]
Wang, X.; Wang, Z.; Ma, X.; Subramanian, S.; Xie, Z.; Shang, C.; Li, X. Analysis of impact toughness scatter in simulated coarse-grained HAZ of E550 grade offshore engineering steel from the aspect of crystallographic structure. *Mater. Charact.* **2018**, *140*, 312–319. [CrossRef]
Hugosson, H.W.; Jansson, U.; Johansson, B.; Eriksson, O. Restricting dislocation movement in transition metal carbides by phase stability tuning. *Science* **2001**, *293*, 2434–2437. [CrossRef]

27. Weaver, J.S.; Li, N.; Mara, N.A.; Jones, D.R.; Cho, H.; Bronkhorst, C.A.; Fensin, S.J.; Gray III, G.T. Slip transmission of high an grain boundaries in body-centered cubic metals: Micropillar compression of pure Ta single and bi-crystals. *Acta Mater.* **2018**, *1* 356–368. [CrossRef]
28. Wang, Z.; Guo, Z.; Shang, C.; Chen, B.; Hui, Y. Characterization of the Stretch Flangeabitity of High-Strength Bainitic Steel: T Significance of Variant Pairs. *Materials* **2021**, *15*, 276. [CrossRef]
29. Wang, Z. Study on the Correlation between Crystallography and Mechanical Properties of High Strength Steels. Ph.D. The University of Science and Technology Beijing, Beijing, China, 2022.

Disclaimer/Publisher's Note: The statements, opinions and data contained in all publications are solely those of the individu author(s) and contributor(s) and not of MDPI and/or the editor(s). MDPI and/or the editor(s) disclaim responsibility for any injury people or property resulting from any ideas, methods, instructions or products referred to in the content.

MDPI AG
Grosspeteranlage 5
4052 Basel
Switzerland
Tel.: +41 61 683 77 34

Metals Editorial Office
E-mail: metals@mdpi.com
www.mdpi.com/journal/metals

Disclaimer/Publisher's Note: The title and front matter of this reprint are at the discretion of the Guest Editors. The publisher is not responsible for their content or any associated concerns. The statements, opinions and data contained in all individual articles are solely those of the individual Editors and contributors and not of MDPI. MDPI disclaims responsibility for any injury to people or property resulting from any ideas, methods, instructions or products referred to in the content.

www.ingramcontent.com/pod-product-compliance
Lightning Source LLC
LaVergne TN
LVHW070446061126
851858LV00011B/173